Distilling Ideas
An Introduction to Mathematical Thinking

Graphs
Groups
Calculus

Mathematics
Through Inquiry

© 2013 by
The Mathematical Association of America (Incorporated)

Library of Congress Control Number: 2013945470

Print ISBN: 978-1-93951-203-1

Electronic ISBN: 978-1-61444-613-2

Printed in the United States of America

Current Printing (last digit):
10 9 8 7 6 5 4 3 2 1

Mathematics Through Inquiry

Distilling Ideas
An Introduction to Mathematical Thinking

Graphs
Groups
Calculus

Brian P. Katz
Augustana College

and

Michael Starbird
The University of Texas at Austin

Published and distributed by
The Mathematical Association of America

Committee on Books
Frank Farris, *Chair*

MAA Textbooks Editorial Board
Zaven A. Karian, *Editor*

Matthias Beck
Richard E. Bedient
Thomas A. Garrity
Charles R. Hampton
John Lorch
Susan F. Pustejovsky
Elsa J. Schaefer
Stanley E. Seltzer
Kay B. Somers

MAA TEXTBOOKS

Bridge to Abstract Mathematics, Ralph W. Oberste-Vorth, Aristides Mouzakitis, and Bonita A. Lawrence

Calculus Deconstructed: A Second Course in First-Year Calculus, Zbigniew H. Nitecki

Combinatorics: A Guided Tour, David R. Mazur

Combinatorics: A Problem Oriented Approach, Daniel A. Marcus

Complex Numbers and Geometry, Liang-shin Hahn

A Course in Mathematical Modeling, Douglas Mooney and Randall Swift

Cryptological Mathematics, Robert Edward Lewand

Differential Geometry and its Applications, John Oprea

Distilling Ideas: An Introduction to Mathematical Thinking, Brian P. Katz and Michael Starbird

Elementary Cryptanalysis, Abraham Sinkov

Elementary Mathematical Models, Dan Kalman

An Episodic History of Mathematics: Mathematical Culture Through Problem Solving, Steven G. Krantz

Essentials of Mathematics, Margie Hale

Field Theory and its Classical Problems, Charles Hadlock

Fourier Series, Rajendra Bhatia

Game Theory and Strategy, Philip D. Straffin

Geometry Revisited, H. S. M. Coxeter and S. L. Greitzer

Graph Theory: A Problem Oriented Approach, Daniel Marcus

Knot Theory, Charles Livingston

Learning Modern Algebra: From Early Attempts to Prove Fermat's Last Theorem, Al Cuoco and and Joseph J. Rotman

Lie Groups: A Problem-Oriented Introduction via Matrix Groups, Harriet Pollatsek

Mathematical Connections: A Companion for Teachers and Others, Al Cuoco

Mathematical Interest Theory, Second Edition, Leslie Jane Federer Vaaler and James W. Daniel

Mathematical Modeling in the Environment, Charles Hadlock

Mathematics for Business Decisions Part 1: Probability and Simulation (electronic textbook), Richard B. Thompson and Christopher G. Lamoureux

Mathematics for Business Decisions Part 2: Calculus and Optimization (electronic textbook), Richard B. Thompson and Christopher G. Lamoureux

Mathematics for Secondary School Teachers, Elizabeth G. Bremigan, Ralph J. Bremigan, and John D. Lorch

The Mathematics of Choice, Ivan Niven

The Mathematics of Games and Gambling, Edward Packel

Math Through the Ages, William Berlinghoff and Fernando Gouvea

Noncommutative Rings, I. N. Herstein

Non-Euclidean Geometry, H. S. M. Coxeter

Number Theory Through Inquiry, David C. Marshall, Edward Odell, and Michael Starbird

A Primer of Real Functions, Ralph P. Boas

A Radical Approach to Lebesgue's Theory of Integration, David M. Bressoud

A Radical Approach to Real Analysis, 2nd edition, David M. Bressoud

Real Infinite Series, Daniel D. Bonar and Michael Khoury, Jr.

Topology Now!, Robert Messer and Philip Straffin

Understanding our Quantitative World, Janet Andersen and Todd Swanson

MAA Service Center
P.O. Box 91112
Washington, DC 20090-1112
1-800-331-1MAA   FAX: 1-301-206-9789

Contents

Preface		**ix**
1	**Introduction**	**1**
	1.1 Proof and Mathematical Inquiry	1
2	**Graphs**	**5**
	2.1 The Königsberg Bridge Problem	5
	2.2 Connections	6
	2.3 Taking a Walk	16
	2.4 Trees	23
	2.5 Planarity	25
	2.6 Euler Characteristic	27
	2.7 Symmetries	32
	2.8 Colorability	34
	2.9 Completing the Walk around Graph Theory	40
3	**Groups**	**43**
	3.1 Examples Lead to Concepts	43
	3.2 Clock-Inspired Groups	52
	3.3 Symmetry Groups of Regular Polygons	55
	3.4 Subgroups, Generators, and Cyclic Groups	56
	3.5 Sizes of Subgroups and Orders of Elements	62
	3.6 Products of Groups	64
	3.7 Symmetric Groups	65
	3.8 Maps between Groups	68

3.9	Normal Subgroups and Quotient Groups	76
3.10	More Examples*	80
3.11	Groups in Action*	81
3.12	The Man Behind the Curtain	85

4 Calculus — **89**

4.1	Perfect Picture	89
4.2	Convergence	91
4.3	Existence of Limits	104
4.4	Continuity	114
4.5	Zeno's Paradox™	123
4.6	Derivatives	127
4.7	Speedometer Movie and Position	135
4.8	Applications of the Definite Integral	137
4.9	Fundamental Theorem of Calculus	141
4.10	From Vague to Precise	145

5 Conclusion — **149**

5.1	Distilling Ideas	149

Annotated Index — **153**

List of Symbols — **165**

About the Authors — **169**

Preface for Students and Instructors

Math textbooks are usually heavy behemoths. They try to provide every argument and example ever wanted. In other words, they hope to be used like an encyclopedia during and after the course. This textbook is gloriously lightweight, even when printed, which is possible because this book has different goals from a traditional book. We hope this text will guide its users to develop the skills, attitudes, and habits of mind of a mathematician.

Of course, it only makes sense to hire a guide if you've got some new territory to explore. Users of this book will become active explorers—experiencing mathematical thinking on their own. What equipment should you bring with you on this exploration? In short, all you need is the desire to work hard and an interest in thinking more clearly. You are not expected to have any previous experience with mathematical proof. If you have some exposure to calculus, then you will already have some experience with mathematical abstraction that will support your exploration, but the content of a calculus course is not needed for this adventure.

This book encourages its users to behave as much like practicing mathematicians as possible. We want students and other people who use this book to learn and to enjoy the process of distilling and exploring ideas. This book helps to embed habits of inquiry in users through the exploration of interesting mathematical content. The book presents a carefully designed sequence of exercises and theorem statements so that its users will be guided to discover both mathematical ideas and also strategies of proofs and strategies of thinking. This preface to students and to instructors explains how the book might be used.

For Students. Mathematics is not a spectator sport. To learn mathematics deeply, it is important to actively grapple with each new idea, putting the pieces into context and seeing connections and meaning. This book presents you with a sequence of challenges that are designed to give you the experience of developing mathematics for yourself. During your work with this book, you will become increasingly proficient at proving theorems on your own and at learning new ideas on your own. You will find that you gain confidence and competence in facing challenging issues.

The strategy of presentation in this book may be a bit unusual to you. You may wonder why we didn't simply write the proofs of the theorems in the book, rather than omitting the proofs and expecting you to figure them out for yourself. Researchers in the field of human learning have found that the best learning occurs through a process of engaged struggle with the ideas. In mathematics, this means constructing proofs. The struggle can be enjoyable along the way, but personally working through ideas lies at the heart of how to learn. And learning how to learn is the real goal of an education.

At first you may be uncertain about what constitutes a mathematical proof and how to even begin to prove something on your own, but after a short while, you will find that you have mental strengths that you did not have before. Like a beginner at tennis, at first it seems impossible to imagine how to strike the ball, much less direct it to where you want it to go. Watching videos of expert tennis players or hearing explanations can only help so much. The basic part of learning a skill is in doing it yourself, including those early steps during which you will inevitably make many mistakes. Just be confident that through your own attempts at doing the exercises and proving the theorems, you will find yourself becoming better and better.

At the very heart of this book is the assumption that you can and will be asked to think about many questions *before* a teacher or classmate tells you how to think about them. This process can be uncomfortable for many students, but it is absolutely critical for making the

Preface   xi

transition from consumer of knowledge to producer of knowledge that is the goal of a college education. In general, you should read the text up to the point of an exercise or theorem and then work on that question, only reading more when you are finished. Many students have had this experience before you, and the exercises and theorem statements are designed to present challenges that will guide you toward increased independent mathematical ability. Try not to get discouraged in the early days when you might well feel a bit confused and overwhelmed. Instead, confidently keep trying and you will become a better thinker.

For Instructors. This book is intended to be used in support of a guided discovery, Inquiry Based Learning method of instruction. When we use the book in our own classes, we ask the students, working either individually or in groups, to do the exercises and prove the theorems and then to present their work to the class; a substantial amount of this work is done by the students in preparation for class. Then we involve the other students by asking them to review part of a proof or to otherwise comment on the presentation. This format of class activity soon gives students the idea that they need to think through ideas on their own. Once they firmly accept that they can personally explore the unknown and can personally determine whether an argument is correct or not, then whole worlds open up to them. They become producers of knowledge rather than consumers of knowledge. Their standards about what they view as understanding rise in this class and other classes; understanding comes to include the ability to explain an idea.

Recent educational research has shown that an early inquiry-based learning experience is particularly beneficial. These kinds of experiences may seem as though they would only work for strong students, but the research also shows that inquiry-based learning experiences are *more* beneficial on average for previously low-achieving students.

If you do not have experience teaching using an Inquiry Based Learning approach, there are many materials about Inquiry Based Learning instruction available, including a mentoring network and workshops. The website for the Academy of Inquiry Based Learning, www.inquirybasedlearning.org, can direct you to many helpful resources. There are many strategies for conducting an IBL course, and the AIBL website can help you to find a method that fits your needs.

The *Through Inquiry* collection of resources is intended to provide instructors with the flexibility to create textbooks supporting a whole range of different courses to fit a variety of instructional needs. To date, the series has four e-units treating, graphs, groups, ε-δ calculus, and number theory. You may combine any collection of these units to create a textbook or textbooks tailored to your needs. Here are some brief descriptions of these units.

- Graphs: This unit asks users to explore the famous ideas flowing from the Königsberg bridge problem, planarity, and colorability. This unit is concrete and visual, which supports students who are new to mathematical proof and abstraction; however, some students find it challenging to be explicit about ideas that are visually obvious to them. It contains several situations that encourage students to generate algorithmic constructions; this experience is particularly effective at getting students to be explicit about their ideas. The theorems in this unit have several applications, which can help motivate those students who are attracted to the usefulness of mathematics.

- Groups: This unit asks users to explore the fundamental algebraic structures of addition, multiplication, and symmetry composition towards a partial classification of groups. This unit is "categorical" in the sense that it leads students to explore sub-objects, product objects, quotient objects, et cetera; as a result, this exploration provides the most clear and explicit list of automatic responses that mathematicians have to new definitions. This unit is quite abstract. It is particularly good at helping students learn to return to the precise language of the definitions. This unit is particularly appealing to those students who love the rigor and abstract beauty of mathematics.

- Calculus: This unit asks users to explore the rigor behind limits, continuity, derivatives, and integrals in ε-δ calculus. This unit con-

Preface

tains calculus content that will probably be familiar to students, which is both helpful and challenging. The students who know calculus will bring intuition about the results, which will allow them to focus on the argumentation; however, the same students are also more likely to try to recall vague explanations from previous courses instead of generating and evaluating new proofs. We believe this unit could also be used with students who do not know any calculus, though it should be taken at a much slower pace, and the students should be asked to work with more examples than are included here. The definitions in this unit contain the long strings of quantifiers that make this topic especially challenging. This unit is particularly appealing to students who intend to teach calculus and to students who seek to understand the mathematical basis of a familiar topic.

- Number Theory: Number theory is a great vehicle for an introduction to proof course. Students are familiar with basic properties of numbers, and yet the further study of number theory leads to fascinating and extremely deep mathematics. This introduction to number theory starts from familiar concepts about integers and leads students to discover the strategies of using definitions, exploring examples, and proving theorems. It treats modular arithmetic, primes, RSA encryption, and additional topics. This extensive unit contains sufficient material for a one semester course plus additional topics for independent study.

Inquiry-based instruction can be used in courses with several different goals. It can be used in introduction-to-proof courses. It can be used to build an alternative entry to the mathematics major, giving students a view of mathematics different from what they have probably seen before, a little more like research. It can be used in a course that gives students who may not be going on in mathematics a sense of some abstract mathematics besides calculus. It is especially valuable for future or current teachers. It is ideal for independent study. And it can be used in more traditional courses to emphasize student-centered learning.

This collection of units contains sufficient material for several semesters of courses, so you will have to choose which units will serve your students for each course. You may choose to select only some of the exercises and theorems in a given unit, particularly if you choose to do more than one unit in a semester. No matter what units you choose and approach you take, you should always leave room for students to ask their own, additional mathematical questions and to state, explore, and prove their own conjectures. Sample, specific threads are described in the Instructor's Resource; however, the main principle is to modify your decisions based on your own students' needs.

Here are some suggestions for which units to use for various purposes.

- **Introduction to proof** – Courses intended to introduce students to proof need to include many opportunities for the students to construct proofs; inquiry-based courses are ideal for providing those opportunities. You could choose to spend the entire term on one topic. Number theory contains enough material for a one semester course; the graph, group, and calculus units would need to be done in great depth or augmented with additional challenges if you choose to treat only one of these topics. Instead, you could choose two or three of these units. We think that graph theory and number theory are particularly good topics for introducing proofs. One advantage of choosing multiple topics for this course is that students get to start fresh with new definitions and see the progress they have made in writing proofs. None of these options assumes that the students have been taught about proof structures. When we use these units for an introduction to proof course, the common proof techniques are distilled through discussion and reflection from the arguments generated by the students.

- **Courses for pre-service/in-service teachers** – An inquiry-based course is particularly important for teachers, because it models a strategy of instructional design; it gives them practice communicating mathematics; it allows them to see the learning process of other students; and it encourages deep understanding of the material that leads to a stable and nuanced understanding. Teachers working with younger students would benefit from a course that uses number theory because numbers are the central concept of el-

Preface

ementary school mathematics. Teachers working with high school students need to understand the basics of abstract algebra and real analysis in order to provide coherence to the high school mathematics. So they would benefit from a course that uses the group theory and calculus units.

- **Alternative entry to the major** – The traditional sequence of mathematics courses for mathematics majors has the unfortunate drawback of avoiding topics in abstract mathematics beyond calculus until later in the students' careers. An alternative is to create a non calculus-based introduction available to students as early as the first year. Selecting topics from the graph, group, and calculus units could provide enticement to further study of mathematics that gives them a taste for a variety of different flavors of mathematics. Such a course is also ideal for science students who would otherwise not see mathematics beyond calculus.

- **Survey for applied math majors** –All mathematics majors should have a working understanding of the basic ideas of algebra and analysis. A course that uses the group theory and calculus units could serve as a survey course treating these two topics.

- **Independent study** – Any unit or collection of units could be used as a resource for a single user or a small group of users who want to learn independently about mathematics. This could be in the form of an independent study under the guidance of a mathematician or a self-guided experience. Since there are many challenging exercises and theorems, learners can do a great deal of work independently. Working through the units provides an excellent quasi-research experience.

- **Core/Topics Courses** – Inquiry-based learning is an effective way for students to learn mathematics. These units represent core and elective topics common in many mathematics departments. An inquiry-based learning experience foster habits of mind, like inquisitiveness and self-efficacy, that are important for all students.

We hope that all users of this book will find it a rich source of enjoyable challenges.

Acknowledgments

We thank the Educational Advancement Foundation and Harry Lucas, Jr. for their generous support of the Inquiry Based Learning Project in general and their support of our work on this project in particular. The Inquiry Based Learning Project has inspired us and has inspired many other faculty members and students across the country. The Educational Advancement Foundation has a clear purpose of fostering methods of teaching that promote independent thinking and student creativity. We hope that this project will contribute to making inquiry based learning methods of instruction broadly available to many faculty members and students nationally. We thank Chris Thoren for his work on some of the images. We also thank the National Science Foundation for its support of this project under NSF-DUE-CCLI grants 0536839 and 0920173. In addition, we would like to thank the Department of Mathematics at The University of Texas at Austin and the Department of Mathematics and Computer Science at Augustana College for creating opportunities in which this work could be carried out.

We would like to thank everyone who has helped this project evolve and become more refined. Most importantly, we would like to thank the many students who have used versions of these units in our classes. They helped us to make the presentation more effective and they energized us by rising to the challenges of personal responsibility, rigor, and curiosity. They inspired us to aim high in education–to realize that mathematics classes can genuinely help students to become better, more creative, more independent thinkers.

1

Introduction

1.1 Proof and Mathematical Inquiry

The word "mathematics" is derived from the Greek term *mathematikos*, meaning "inclined to learn." Many people associate mathematics with computation and with abstraction, but those characteristics are not the attributes that make mathematicians so sought-after on the job market and so effective at grappling with difficult issues in the world. Instead, mathematics is especially powerful because of the potent ways it asks questions, distills concepts, and explores the unknown: mathematicians are "universal learners," trained to employ certain skills in response to any new idea.

This book is designed to convey the habits of *proof* and *mathematical inquiry*, that is, the inclination and the skill to ask and explore questions. You will build that powerful skill by personally engaging in mathematical inquiry on your own, guided by the challenges given in this book.

Proof. Every academic discipline has a particular mode of argumentation, and mathematical argumentation is called proof. The word "proof" is used outside of mathematics, but the two usages have slightly different meanings. Outside of mathematics, proof usually refers to evidence or the data that supports a claim. A fingerprint is seen as proof that a suspect has been at a certain location, but anyone who has seen an episode of *Bones* knows that there are ways to misuse this kind of proof.

In mathematics, a proof is simply an argument in which every statement follows logically from the previous statements, definitions, and assumptions and only from these preceding ideas. As a result, many proofs are born from close work with precise definitions instead of intuitive understandings of the terms involved.

This book asks you to construct your own proofs of the challenges posed and then spend time honing your proof skills by sharing the results with other inquirers. With this book, you are far more than a "reader." Your own active work will result in your learning both the mathematical content and the skills of mathematical inquiry.

The statements to be proved come in many varieties, but the four most common are Theorem, Lemma, Corollary, and Exercise. A theorem is a general statement that asserts that a specified conclusion always is valid if the stated assumptions are true. A lemma is a theorem that is intended to be used immediately to help prove a following theorem. A corollary is a theorem whose proof is almost entirely accomplished by applying the preceding theorem. And an exercise is, in a sense, a theorem about a very specific situation. And a Theorem* represents a challenge. Regardless of the name, each of these statements needs a proof; even an exercise that can be "answered" by a number or a drawing should be accompanied by an explanation (i.e., a proof).

Because a proof is an argument, it is made up of complete sentences, including verbs in each sentence. The word "equals," often written as "=," *is* a verb, but most of the proofs you will create will contain far more words than symbols.

We agree with Albert Einstein when he said that the whole of mathematics (and science) is merely a refinement of everyday thinking. Proving theorems in not a different way of thinking—it is merely a refinement of clear thinking. All people naturally develop the beginnings of the skill of constructing proofs, so this book is organized to provide you opportunities to nurture that skill in a variety of contexts with gradually increasing complexity, difficulty, and ab-

1.1. Proof and Mathematical Inquiry 3

straction. Since we strive to convey the idea that proofs are merely refinements of everyday thinking, we resist the urge to give you formulaic proof templates, which would tend to suggest that your task is to mimic some magical incantation. Through working through this book, you should expect to develop an understanding of mathematical knowledge and of proof so that you are empowered to use that added strength in all your future mathematical experiences and beyond.

Mathematical Inquiry. Most users of this book have little experience with mathematical inquiry, that is, with asking their own mathematical questions, creating new mathematical ideas, and exploring mathematical concepts. This book is intended to give you those experiences. As you work through this book, you will add new, powerful inquiry skills to your tool belt: abstraction, exploration, conjecture, justification, application, and extension.

The journey of discovery often starts with a specific challenge or puzzle. Abstraction is the step of isolating essential ingredients in a complex or subtle situation and pinning it down to create definitions and insights. Exploration is the process of generating examples and looking for patterns, seeking the full range of possibilities of this abstract idea. Conjectures articulate the apparent patterns. Justification is the process of proving the conjectures (creating theorems) or disproving the conjectures. Applications of our new theorems often give us insight into our original challenge or puzzle. Extension starts the inquiry process again with new challenges that have become available to us because of our new insights.

The process of creating abstract mathematics is accomplished through practicing these potent strategies of inquiry. While working through this book, you will develop and distill your own experiences into an understanding of mathematical inquiry.

Exploring mathematical ideas is an active process. You will not understand mathematical thought unless you personally participate in mathematical investigations. So this book actually is an invitation to you to think through the development of various mathematical concepts with the aid of our guidance. We have tried to design the experience to maximize the satisfaction you will feel in making mathematical ideas your own. In addition, we have purposefully left many questions unasked to give you room to explore your own interests as you proceed. Ideally, every time you work with these ideas, you will have a question that is neither asked nor answered in this book.

This book strives to help you see the wonder of mathematical exploration and to become personally proficient at mathematical inquiry and proof. Having these skills is the real goal of a mathematics education. We hope you will leave more "inclined to learn" than you were when you arrived.

2

Graphs

2.1 The Königsberg Bridge Problem

Turn back the clock to the early 1700's and imagine yourself in the city of Königsberg, East Prussia. Königsberg was nestled on an island and on the surrounding banks at the confluence of two rivers. Seven bridges spanned the rivers as pictured below.

One day, Königsberg resident Friedrich ran into his friend Otto at the local Sternbuck's coffee shop. Otto bet Friedrich a Venti Raspberry Mocha Cappucino that Friedrich could not leave the café, walk over all seven bridges without crossing over the same bridge twice (without swimming or flying), and return to the café. Friedrich set out, but he never returned.

The problem of whether it is possible to walk over all seven bridges without crossing over the same bridge twice became known as the

Königsberg Bridge Problem. As far as we know, Friedrich is still traipsing around the bridges of Königsberg, but a mathematician named Leonhard Euler did solve the Königsberg Bridge Problem in 1736, and his solution led to the modern area of mathematics known as *graph theory*.

2.2 Connections

One of the richest methods for developing mathematical ideas is to start with one or more specific problems and pare them down to their essentials. As we isolate the essential issues in specific problems, we create techniques and concepts that often have much wider applicability.

Sometimes it's quite hard to isolate the essential information from a single problem. If we consider several problems that "feel" similar, often the feeling of similarity guides us to the essential ingredients. It's a little like how, when playing the games *Catch Phrase* or *Taboo*, you choose several other words that have the secret word as a common thread. This process is important in creating the subject of *graph theory*. So let's begin by considering several additional questions that feel similar to the Königsberg Bridge Problem.

As you read the following questions for the first time, instead of trying to solve them, think about what features of each question are essential and look for similarities among the questions.

The Paperperson's Puzzle. Phlipper, the paperperson, has a paper route in a residential area. Each morning at 5:00 am a pile of papers is delivered to a corner in her neighborhood pictured below.

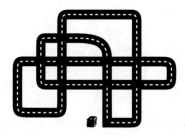

2.2. Connections

She puts all the papers in the basket of her bike and rides around the neighborhood flipping the papers in the general direction of the subscribers' houses. She rides down the middle of the streets and throws papers on both sides as she goes. When she finishes her route, she returns the leftover papers to the same location from which she started. The question is whether she can complete her route without having to ride over the same street more than once.

The Königsberg Bridge Problem and the Paperperson's Puzzle have the similarity of taking a journey and returning to the starting point. However, some additional questions have similarities even though they do not involve motion.

The Gas-Water-Electricity Dilemma. Three new houses have just been built in Houseville, and they all need natural gas, water, and electricity lines, each of which is supplied by a different company as pictured below.

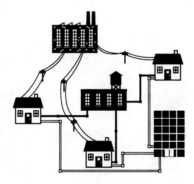

Can each utility company lay a utility line to each house without having any of the utility lines cross?

The Five Station Quandary. Casey Jones wanted to build an elaborate model train set. He set up five stations and wanted to run tracks that connected each station directly to every other station. Could he build his layout with no crossing tracks, bridges, or shared routes?

***Exercise* 2.1.** Before you try to solve these problems, stop for a moment. What features of these problems are similar to one another? *Do* *not go on* until you think of at least one similarity among the problems.

Story problems are the bane of existence to non-mathematically oriented people, but mathematicians know exactly how to begin, namely, *abstraction*; that is, to isolate the salient information and to ignore the irrelevant information. The abstracted concepts and techniques we create will not only help us solve these problems but will also be applicable to any other problem whose abstracted essence is the same.

Here we will discuss the strategy of abstraction in the context of the Königsberg Bridge Problem, but please take analogous steps for the other problems as well.

In the Königsberg Bridge Problem, what is important about the picture of the city? Does it matter how big the island is? Does it matter how long any of the bridges are? Does it matter that there are two bridges between the northwest sector of town and the island? Ask yourself, "Which features of the problem set-up are relevant, and which features are not?". Asking yourself these questions is a big step towards mathematical maturity, and helping you to adopt the habit of asking yourself effective questions is one of the major goals of this book.

The features that seem to matter for the Königsberg Bridge Problem are the different locations (three land masses and the island) and the different bridges that cross between pairs of those locations. So one way to abstract the essence of this situation is to draw a dot for each location and a (possibly curved) line segment or edge for each bridge that connects a pair of dots (land masses). Since the problem does not ask about distance, the abstraction need not attempt to reflect any of the distances involved. Similarly, the physical locations of the land masses do not affect the problem, so the dots do not need to be positioned in any way that reflects the original city layout. The essential ingredients are locations and connections.

***Exercise* 2.2**. Draw an abstracted picture that corresponds to the Königs-berg Bridge Problem. Your picture should consist of dots and lines. Explain in your own words why this is a good representation.

2.2. Connections

Exercise **2.3**. Draw similar abstracted pictures for each of the other challenges described above, the Paperperson's Puzzle, the Gas-Water-Electricity Dilemma, and the Five Station Quandary. Your pictures should consist of dots and lines. In some cases, you must face the issue that you may not be able to draw all the lines connecting the dots without having the lines cross on the page. You will need to devise a strategy for indicating when an intersection of lines in your representation really shouldn't be there. Explain in your own words why these are good representations. In each case, what do your dots represent? What do your lines represent?

Exercise **2.4**. Attempt to solve the Five Station Quandary, that is, attempt to draw connecting tracks between every pair of stations without any tracks crossing each other. If you cannot accomplish this solution, draw as many connecting tracks as you can without crossing and then draw in any remaining tracking indicating places where you would need a bridge or tunnel to avoid unwanted intersections. Develop a notation to indicate a bridge or tunnel.

Perhaps you can think of alternative ways to abstract the essence of the situation that does not use a picture at all and, hence, altogether avoids the issue of unwanted intersections of lines.

Exercise **2.5**. Describe a new system for representing these situations that does not involve dots and lines but still contains the same information about the connections. Represent the data of one of the challenges above in your new system.

All of these problems resulted in abstractions that have similar characteristics. The visual representations all had dots and lines where each line connected two dots. Your non-visual representations probably used letters to represent locations or houses and utilities or people, while connections between pairs were indicated somehow, perhaps by writing down the pairs of letters that had a connection. Both the visual representation and the written one contained the same information and that information has two basic ingredients—some things and some connections between pairs of things. Once we have isolated these ingredients, we are ready to take an important step in the development of our concept, and that is to make some *definitions*.

Notice that we didn't start with the definitions. This process is typical of mathematical invention: we explore one or more situations that contain some intuitive or vague ideas in common and then we pin down those ideas by making a formal definition. Definitions are a mathematician's life's blood because they allow us to be completely clear about what is important and what is not important in a statement.

In all our examples above, including the Königsberg bridges, the train tracks, and the utilities and houses, we isolated the important features as things and connections. So we are ready to make a definition that captures situations of that type. The word we use to capture this abstract situation is a *graph*. Finally, here is our definition. It is very abstract. It says basically that if we start with any collection (a set) and then a bunch of pairs of those things (that is, which pairs of those things are connected), then we have a graph. It will take some getting used to before this completely abstract definition makes sense, but by looking at examples and proving theorems about graphs, they will become familiar and natural.

***Definition* 2.1.** Let $V = \{v_1, v_2, v_3, \ldots, v_n\}$ be a finite, nonempty set, and let $E = \{e_1, e_2, e_3, \ldots, e_m\}$ be a set where each e_i is a pair of elements of the form $\{v, v'\}$, where v and v' are in V. Then the pair (V, E) is called a **graph**. We call elements of V the **vertices** and elements of E the **edges**. Sometimes we will write $G = (V, E)$ and call the graph G.

When thinking about vertices, think about locations like the different locations (land masses and the island) in Königsberg, and when thinking about edges remember the bridges, each of which connected some pair of locations in the city. Alternatively, think of vertices as the train stations and think of each edge as the track between them.

Notice that our abstract definition of a graph does not overtly have a visual component. However, we could make an object that corresponds to a graph by taking a pebble for each vertex and connecting a pair of pebbles with a piece of string for each edge of the graph. Or we could draw a picture that corresponds to a graph by drawing a dot for each vertex and drawing a possibly curved line segment connecting a pair of vertices for each edge of the graph. As before, in our drawing of a graph, we would have to make certain that our representation

2.2. Connections

clearly showed that edges do not intersect one another. Each edge is separate.

The word "graph" comes from the Greek root word meaning "to write". In high school math classes, "to graph" means "to draw", as in "graphing a function". Perhaps we should have chosen a different name since a graph is not inherently visual, but the term is too firmly entrenched to change now, and often an appropriate visual representation of a graph gives us valuable insights.

We can be somewhat satisfied with our definition, but now we have to step back and ask ourselves whether there are any issues that need to be addressed. If we look at the graph corresponding to the Königsberg Bridge Problem, we might notice a potential issue, namely, there is a pair of land masses that are connected with two different bridges. In fact, there are two such pairs of land masses. In terms of the abstract definition of a graph, that means that the same pair of vertices appears as distinct edges in E. We have isolated an issue. So let's explicitly allow E to contain multiple copies of a pair $\{v, v'\}$, just as we allowed multiple bridges between the same land masses in the Königsberg Bridge Problem. So we will allow multiple edges between the same pair of vertices and indicate their presence by writing the same pair down as many times as there are multiple edges between those vertices. After we have isolated the idea of multiple edges, we can define graphs with that feature.

Definition 2.2. A graph $G = (V, E)$ is said to have **multiple edges** if E contains two (or more) distinct copies of an edge $\{v, v'\}$. In plain language, G has multiple edges if it has two vertices that are connected by more than one edge. Technically, the existence of multiple edges connecting the same two vertices means that E is a multiset, not a set, but we will ignore this issue.

We've gotten pretty abstract, pretty quickly. The following exercise is to make sure you're following.

Exercise 2.6. Imagine a dinner party at which each of the participants shakes the hand of some of the other participants. Show how to create a graph that represents the situation. Let the vertices correspond to people and the edges correspond to handshakes. What would it mean for this graph to have multiple edges? What would it mean for one vertex to have no edges?

Another issue that comes to mind is whether the edge $\{v, v'\}$ is the same or different from the edge $\{v', v\}$. That is, does the order of the vertices in an edge make a difference? Well, we could choose either answer. In the situations that generated our concept, the order did not matter (we could walk over the bridges in either direction and the handshakes did not have a direction to them, for example), so we will choose not to distinguish between $\{v, v'\}$ and $\{v', v\}$. So for this concept of a graph, we could replace any of our edges with a pair of vertices in the opposite order and say that that is the same graph.

If we chose to view differently ordered edges as different, then we would be describing something that is referred to as a *directed graph*. Directed graphs would be appropriate for capturing some other situations. For example, suppose there were one-way signs on the bridges of Königsberg, then a directed graph would be required to capture the restrictions that the new problem presented. Directed graphs also make more sense when modeling the spread of a disease, since we would want the representation to capture the idea that an infected person infects a non-infected person. Similarly, the evolutionary tree of life is well represented by a directed graph.

We have yet one more issue that we may want to make a decision about: should we allow an edge to go from a vertex to itself? None of our generating scenarios has such a situation; however, we could easily imagine such a situation. We could imagine a bridge that starts and ends on the same land mass, like an overpass, for example. So we will choose to allow edges of the form $\{v, v\}$. Since that edge is rather distinctive looking, we will give it a name.

Definition 2.3. Let $G = (V, E)$ be a graph with a vertex v. Then an edge of the form $\{v, v\}$ is called a **loop** (at v).

Now let's get accustomed to the vocabulary of a graph by looking at the Königsberg Bridge Problem in our new terms.

***Exercise* 2.7.** Carefully, using the definitions we have just chosen, construct a graph for the Königsberg Bridge Problem, $K = (V, E)$. Give each vertex a label (probably just a lower-case letter); then, using

2.2. Connections

these labels, write V and E for this graph.

***Exercise* 2.8.** For each of the other challenges, think about how you would specify a graph $G = (V, E)$. It would be tedious to do all of them. So pick at least one to write out carefully.

In thinking about the Königsberg Bridge Problem, it would be reasonable to say that a bridge "has endpoints v_1 and v_2" where $\{v_1, v_2\}$ was an edge in the graph. So there is some natural vocabulary that will help us to discuss questions about graphs.

***Definition* 2.4.** Let G be a graph containing vertices v and v' and the edge $e = \{v, v'\}$. Then e has **endpoints** v and v', and v and v' are **adjacent by** e.

Making this definition lets us use some more intuitive and familiar language to talk about graphs. In particular, now a graph has multiple edges if it has a pair of vertices that are the endpoints of two distinct edges; two loops at the same vertex count as multiple edges, but a single loop does not. However, there are some weird side effects too: if G contains the loop $\{v, v\}$, then v is adjacent to itself. If we're going to go to all the trouble to carefully create definitions, then we must also be careful when using common language to talk about the ideas.

When we look at our visual representations of the Königsberg Bridge Problem, the Paperperson's Puzzle, and the Gas-Water-Electricity Dilemma, one feature that we see in describing those graphs concerns the number of edges that emerge from each vertex.

***Definition* 2.5.** If v is a vertex, then we define the **degree** of v, written as $\deg(v)$, to be the number of edges with an endpoint v, where a loop counts twice. The **total degree** of a graph G is the sum of the degrees of the vertices of G.

***Exercise* 2.9.** 1. Given a graph $G = (V, E)$, describe a procedure for computing the degrees of the vertices *without* drawing a picture of the graph.

2. Compute the degrees of the vertices in the Königsberg Bridge Problem using the procedure you described in the previous part of this exercise; make sure those answers agree with the numbers you get by just looking at your visual representation.

3. Write out a specific example of a graph with at least five vertices and compute the degree of each vertex and the total degree of the graph.

The next theorem is a general statement, distilled from our experience with examples. Your job, as with all such theorem statements, is to provide a proof of the statement, that is, an ironclad reason that the statement is true.

***Theorem* 2.10.** The total degree of any graph is even.

The next statement is called a corollary, which is just a theorem that follows directly as a consequence of a previous theorem (so you should prove corollaries as well).

***Corollary* 2.11.** Let G be a graph. Then the number of vertices in G with odd degree is even.

One of your new habits as a mathematician is to check how theorems work in particular cases every time you do a proof. This habit helps to make abstract mathematics meaningful.

***Exercise* 2.12.** Confirm the truth of the theorem and corollary above in the Königsberg Bridge Problem graph and in the graph you constructed in part 3 of Exercise **2.9**.

Theorem **2.10** and its corollary point out restrictions on graphs with respect to the degrees of vertices. These insights allow us to determine whether graphs could exist with various properties.

***Exercise* 2.13.** Determine whether the following data could represent a graph. For each data set that *can* represent a graph, determine *all* the possible graphs that it could be, describe each graph using pictures and set notation, and explain why your list is complete. If no graph can exist with the given properties, state why not.

1. $V = \{v, w, x, y\}$ with $\deg(v) = 2$, $\deg(w) = 1$, $\deg(x) = 5$, $\deg(y) = 0$

2. $V = \{a, b, c, d\}$ with $\deg(a) = 1$, $\deg(b) = 4$, $\deg(c) = 2$, $\deg(d) = 2$

3. $V = \{v_1, v_2, v_3, v_4\}$ with $\deg(v_1) = 1$, $\deg(v_2) = 3$, $\deg(v_3) = 2$, $\deg(v_4) = 5$

2.2. Connections

You proved earlier that the total degree of any graph is even. Let's consider a sort of converse question, namely, if the degrees of vertices are given such that the total degree *is* even, can we create a corresponding graph?

***Exercise* 2.14.** If you are given a finite set V and a nonnegative integer for each element in the set such that the sum of these integers is even, can V be realized as the vertices of a graph with the associated degrees? If so, prove it. If not, give a counter-example.

The set-up of the Königsberg Bridge Problem was well modeled by the concept of a graph, which you have drawn. After abstracting the set-up, we must also translate the challenge of the problem in terms of the associated graph. In posing the Königsberg Bridge Problem, Otto was asking whether it is possible to trace every edge (bridge) of the graph without picking up his pencil and without going over any edge more than once.

***Exercise* 2.15.** Try to trace your Königsberg Bridge graph without picking up your pencil and without going over any edge more than once. You can put the Sternbucks anywhere you like; try several locations. Does the starting place affect the answer?

If we can trace one visual representation of the Königsberg Bridge graph, we can trace any correct representation, which is why we can abuse language and talk about *the* (visual representation of the) graph when working on this problem. But to avoid this subtlety entirely, we can ask Otto's question about our graph where the graph is presented in the set notation $K = (V, E)$.

***Exercise* 2.16.** Translate the Königsberg Bridge Problem into a question about its graph, $K = (V, E)$, without reference to a visual representation of K.

The Königsberg Bridge Problem was modeled by a graph, and its challenge was described in terms of a tracing problem. This problem naturally encourages us to explore the general question of when we can trace a graph without picking up our pencil and without going over any edge more than once.

***Exercise* 2.17.** Draw the graph associated with the Paperperson's Puzzle. Try to trace the graph without picking up your pencil and without going over any edge more than once.

To gain further experience with this traceability question, an excellent strategy is to try many example graphs and observe which ones seem to be traceable and which ones seem not to be traceable. Then we can try to isolate what features of a graph seem to make it traceable. For now, we will say that a graph is **traceable** if the edges can be lined up like dominoes, with matching ends, using each edge exactly once; in other words, a graph is traceable if you can draw it without picking up your pencil or repeating edges. If the ordering of the edges has the same starting as ending point, then the graph is traceable while returning to the start.

***Exercise* 2.18**. For each graph pictured below, try to trace the graph without picking up your pencil and without going over any edge more than once. Look for some feature or features among the graphs that distinguish those you can trace from the ones that you can't. You may not be able to characterize those graphs that are traceable, but perhaps you can isolate some features of a graph that definitely make it traceable or definitely make it untraceable.

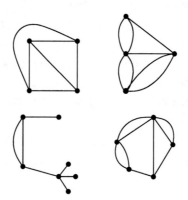

2.3 Taking a Walk

Looking at examples is a great way to begin to explore an idea, but at some point it is valuable to become a bit more systematic in the investigation. Starting with simple cases is an excellent strategy for developing insight. So let's consider some simple graphs to see whether

2.3. Taking a Walk

we can discover some sort of pattern among those that are traceable or untraceable.

If there were only one bridge between two land masses, then the edge could be traced, but it would be impossible to return to the starting place without retracing the same edge. Recall that in the Königsberg Bridge Problem, Otto challenged Friedrich to return to his starting place, so we must consider that restriction. If only one bridge existed and it connected the same island to itself, then we could traverse the bridge while starting and ending at the same point. That is, if a graph had only one edge and that edge were a loop, then we could trace the graph returning to where we started.

Now let's consider graphs with two or three edges.

***Exercise* 2.19**. Draw all possible graphs that contain two or three edges. Argue that your list is complete. Which are traceable while returning to the start, which are traceable, and which are not traceable?

Now investigate graphs with four edges.

***Exercise* 2.20**. Draw all graphs with four edges without loops or vertices with degree 0. Argue that your list is complete. Which graphs with four edges are traceable (with and without returning to the start)? Try to be systematic and try to isolate some principles that seem pertinent to traceability.

Perhaps you will observe that the degrees of the vertices are important for the issue of traceability.

***Exercise* 2.21**. For each of the graphs you drew in Exercises **2.19** and **2.20** as well as those for the Königsberg Bridge Problem and the Paperperson's Puzzle, make a chart that records the degree of each vertex. Do you see something that separates the good from the bad (traceable from not traceable)?

We translated Otto's Königsberg Bridge Problem into a question about graph theory, and now we will formalize what it means to find a solution. Just as making our definitions helped us decide what was important about the problem, formalizing the question helps us see how to break it down into more manageable steps.

The act of tracing the edges of a graph is a fairly clear process, but there are really several different ways of moving about a graph, some involving the proviso of not repeating edges and the more basic idea of just moving around. So let's take the step of pinning down some definitions about how we can move about on a graph. The first definition refers to moving from one vertex to another, but there is no restriction about repeating the same edge.

Definition 2.6. Let G be a graph with vertices v and w. A **walk** from v to w, W, is a finite sequence of adjacent vertices and edges of G of the form

$$W: v(= v_0), e_1, v_1, e_2, v_2, e_3, \ldots, v_{k-1}, e_k, w(= v_k)$$

where the v_i's are vertices of G, and for each i, e_i is the edge $\{v_{i-1}, v_i\}$. We explicitly allow a trivial walk from v to v, $T : v$, which is just one vertex without any edges.

Walks that do not repeat an edge are of special interest, for example in the Königsberg Bridge Problem, so it is good to observe that if there is a walk between two vertices, there is a walk that has no repeated edges.

Theorem 2.22. Let G be a graph that has a walk between vertices v and w. Then G has a walk between vertices v and w that does not use repeated edges.

The next theorem states that a walk with no repeated vertices cannot have repeated edges either.

Theorem 2.23. Let G be a graph and $W: v_0, e_1, \ldots, e_n, v_n$ a walk in G such that the vertices v_i are all distinct. Then W has no repeated edges.

One basic question we can ask about a graph is whether we can walk from one vertex to another.

2.3. Taking a Walk

***Definition* 2.7.** Let G be a graph with vertices v and w.

1. We say v **is connected to** w if there exists a walk from v to w.

2. The graph G is **connected** if every pair of vertices of G is connected. If not, we say G is **disconnected**.

***Exercise* 2.24.** Show that your graph of the Königsberg Bridge Problem is connected. Carefully use the definitions. Also, give an example of a graph that is not connected.

It is obvious, visually, when a graph is connected, at least when it has a small number of vertices, but that is different from a proof. As the last exercise hopefully showed you, there's a lot to write down to show that a graph is connected. The following theorem helps shorten the work; it also tells us that the term "connected" behaves as we use it in common English.

***Theorem* 2.25.** Let G be a graph with vertices u, v, and w.

1. The vertex v is connected to itself.

2. If u is connected to v and v is connected to w, then u is connected to w.

3. If v is connected to w, then w is connected to v.

In other words, "connected" is an *equivalence relation.*

In the statement of the Königsberg Bridge Problem, recall that Otto's challenge involved returning to the starting place. Notice that in our original definition of a walk, the beginning and ending vertices had no restrictions, so they could actually have been the same vertex. So now we define a special walk that does start and end at the same vertex and has no repeated edges.

***Definition* 2.8.** A **circuit** is a walk with at least one edge that begins and ends at the same vertex and never uses the same edge twice.

***Exercise* 2.26.** Write out some of the circuits in $K = (V, E)$, the Königsberg Bridge Problem graph. Find at least one circuit that contains at least one repeated vertex (other than the initial/final vertex).

When we have described a mathematical entity, such as a graph, it is often useful to look at smaller such objects that are contained in it. This strategy leads to the concept of a subgraph.

***Definition* 2.9**. Let $G = (V, E)$ and $G' = (V', E')$ be graphs. If $V' \subset V$ and $E' \subset E$, then we say that G' is a **subgraph** of G.

Much like checking a theorem in special cases to understand its meaning more thoroughly, mathematicians observe definitions in examples to help them understand the definitions.

***Exercise* 2.27**. Consider the graph G below. Choose several sets of three vertices from this graph and draw all subgraphs of G with exactly those three vertices.

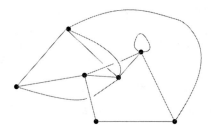

The next theorem observes that removing an edge from a circuit in a connected graph will not disconnect the graph.

***Theorem* 2.28**. Let $G = (V, E)$ be a connected graph that contains a circuit C. If e is an edge in C, then the subgraph $G' = (V, E \setminus \{e\})$ is still connected.

Note that a circuit in a graph is *not* a subgraph; however, the set of vertices and the set of edges in the circuit do form a subgraph.

When talking about a vertex in a graph and also in a subgraph, confusion could arise about the vertex's degree. If G' is a subgraph of G, and a vertex, v, is in both G and G', then we will use the notation $\deg_G(v)$ and $\deg_{G'}(v)$ to denote its degree in G and G' respectively.

***Theorem* 2.29**. Let G be a graph. Let C be a subgraph of G that consists of the vertices and edges that belong to a circuit in G. Then $\deg_C(v)$ is even for every vertex, v, of C.

2.3. Taking a Walk

The Königsberg Bridge Problem produced a graph that we sought to traverse without lifting our pencil or repeating an edge. So that problem gives rise to a definition that captures this kind of traceability.

***Definition* 2.10**. Let G be a graph. An **Euler circuit** for G is a circuit in G that contains every vertex and every edge of G.

Our definition of an Euler circuit was motivated by trying to capture ideas suggested in the Königsberg Bridge Problem. So let's see whether we can restate that puzzle using our new vocabulary.

***Exercise* 2.30**. Restate the Königsberg Bridge Problem using our formal definitions.

In some sense, restating a question in formal terms does not make any progress towards solving it; however, such a restatement can be helpful. Now we are clear on what we seek to find in our graph: we seek a walk with various restrictions.

One natural attempt to solve the Königsberg Bridge Problem would be simply to start walking without going over the same bridge twice and to continue as far as possible. When you can go no further without repeating a bridge, you can make an observation about where you end up.

We are now ready to characterize those graphs that have an Euler circuit. Determining whether a graph has an Euler circuit turns out to be easy to check. It is always a pleasure when a property that appears to be difficult to determine actually is rather simple.

***Euler Circuit Theorem* 2.31**. A graph G has an Euler circuit if and only if it is connected and every vertex in G has even, positive degree.

If you truly understand the proof of this theorem, you should be able to take a graph and produce an Euler circuit, if it has one, using the technique implicit in your proof. So here is an exercise that lets you explore the method of the proof rather than just the statement of it.

***Exercise* 2.32**. In the following graphs, find an Euler circuit using a method that successfully proved the Euler Circuit Theorem.

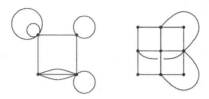

We can now definitively complete the Königsberg Bridge Problem by translating our solution back into the language of bridges and locations. This application represents the culmination of an important pattern in how abstract mathematics can be used to solve problems in the real world. We began in the real world, distilled real features into mathematical ideas, solved the problem in the abstract domain, and now are returning back to the problem in the real world.

***Exercise* 2.33.** Solve the Königsberg Bridge Problem. Write your solution in a way that Otto could understand from start to finish, that is, write your answer thoroughly in ordinary English, or Old Prussian, if you prefer.

Similarly, we can settle the Paperperson's Puzzle.

***Exercise* 2.34.** Solve the Paperperson's Puzzle.

We've finished with Otto's challenge and the Königsberg Bridge Problem, but now we need to think about what other kinds of theorems are true. The first place to look for new theorems is in modifying theorems we've already proved. The second place to look is back at the actual proofs we've produced; sometimes when looking back and summarizing an old proof we realize that simply changing the hypotheses would produce new theorems or that we've actually proved something more than we set out to show.

In that vein, we can ask: Under what circumstances can we trace a graph if we don't have to end where we started?

***Definition* 2.11.** An **Euler path** for a graph G is a walk from a vertex v to a vertex w in G that contains every vertex of G and contains every edge of G exactly once.

***Euler Path Theorem* 2.35.** A graph G has an Euler path if and only if G is connected and has zero or two vertices of odd degree and all other vertices have even, positive degree.

2.4. Trees

Let's make certain that the distinction between an Euler path and an Euler circuit is clear.

***Exercise* 2.36.** Give an example of a graph with an Euler path but not an Euler circuit. What must be true of any such example?

Again, let's practice the method of proof for the Euler Path Theorem.

***Exercise* 2.37.** In the following graphs, find an Euler path using a method that successfully proved the Euler Path Theorem.

2.4 Trees

We've proved a large number of theorems about graphs with circuits and when graphs have certain kinds of circuits. We now turn our attention to some interesting theorems about graphs without circuits, *trees*.

***Definition* 2.12.** A graph is called a **tree** if it is connected and has no circuits.

***Exercise* 2.38.** A start-up airline, AirCheap, only flies to four cities, and all flights go through Wichita. But from Wichita you can fly to Austin, Denver, or Chicago. Construct a graph that has vertices corresponding to the cities and edges corresponding to flights for AirCheap. Is the graph a tree? Justify your answer.

One feature of a tree is that it must contain vertices with low degree. Vertices with degree one are sometimes called **leaves**.

***Theorem* 2.39.** Any tree that has more than one vertex has a vertex of degree one, in fact, it has at least two vertices of degree one.

When we actually look at a tree, we notice that there are often quite a few vertices of degree one (leaves). This observation invites us to explore the question of how the number of degree one vertices relates to other features of the tree.

***Exercise* 2.40.** By drawing a few examples, explore the relationship between the number of degree one vertices of a tree and other features of the tree. Make a conjecture and prove it.

We can tell whether a graph is a tree simply by comparing the number of its vertices with the number of its edges.

***Exercise* 2.41.** There is a simple relationship between the number of vertices and edges in a tree. Make a conjecture of the following form and prove it: A graph with n vertices is a tree if and only if G is connected and has _____ edges.

Trees are particularly simple examples of graphs. In graphs with circuits, there are often many different ways to get from one vertex to another, but in a tree, there is only one option.

***Theorem* 2.42.** If v and w are distinct vertices of a tree G, then there is a unique walk with no repeated edges in G from v to w.

The previous theorem implies that trees are disconnected by the removal of any edge.

***Corollary* 2.43.** Suppose $G = (V, E)$ is a tree and e is an edge in E. Then the subgraph $G' = (V, E \setminus \{e\})$ is not connected.

Every graph has subgraphs that are trees, called subtrees. Connected graphs have subtrees that contain all the vertices of the graph. Sometimes we can use these subtrees as starting points for analyzing the larger graph.

***Theorem* 2.44.** Let G be a connected graph. Then there is a subtree, T, of G that contains every vertex of G.

***Definition* 2.13.** Let G be a graph. Then a subtree T of G is a **maximal tree** if and only if for any edge of G not in T, adding it to T produces a subgraph that is not a tree. More formally, a subtree

2.5. Planarity

$T = (W, F)$ of $G = (V, E)$ is a **maximal tree** of G if and only if for any $e = \{v, w\}$ in $E \setminus F$, $T' = (W \cup \{v\} \cup \{w\}, F \cup \{e\})$ is not a tree.

Theorem 2.45. A subtree T in a connected graph G is a maximal tree if and only if T contains every vertex of G.

2.5 Planarity

Earlier, we ran across the issue of whether we could draw a graph in the plane without having edges cross. If a graph can be drawn without edges crossing, we can often use geometric insights to deduce features about the graph. In the next two sections, we investigate issues concerning graphs that can be drawn in the plane without edges crossing.

Definition 2.14. A graph G is called **planar** if it can be drawn in the plane ($\mathbb{R}^2$) such that the edges only intersect at vertices of G.

Our first observation is that trees can always be drawn in the plane.

Theorem 2.46. Let G be a tree. Then G is planar.

Remember that a graph is just two sets $G = (V, E)$. When a graph is presented in this formal way, it is far from obvious whether the graph is planar. To aid in our exploration of planarity, let's describe some new families of graphs.

Definition 2.15. 1. For a positive integer n, the **complete graph on n vertices**, written K_n, is the graph having n vertices, containing no loops and containing a unique edge for each pair of distinct vertices.

2. For positive integers m and n, the **complete bipartite graph**, $K_{m,n}$, is the graph with $m + n$ vertices, $\{v_1, v_2, \ldots, v_m\}$ and $\{w_1, w_2, \ldots, w_n\}$, where each v_i is connected to each w_j by a unique edge and having no loops or other edges.

Exercise 2.47. Draw graphs of K_3, K_4, K_5, $K_{2,3}$, $K_{3,3}$, and $K_{2,4}$. Which appear to be planar graphs? Are any of them familiar?

The statements in the next exercise are quite hard to prove rigorously. Showing that something is planar only requires finding one particular way to draw it; but showing that something is *not* planar involves showing that no arrangement is possible.

***Exercise* 2.48.** The following statements are true theorems. But, instead of trying to find ironclad proofs now, give informal, plausible arguments that they are true.

1. The graph $K_{3,3}$ is not planar.

2. The graph K_5 is not planar.

Later we will be in a position to give more rigorous proofs.

Even if we can't prove these statements, we can interpret their consequences. Recall two of our motivating questions for graph theory.

- The Gas-Water-Electricity Dilemma: Three new houses have just been built in Houseville, and they all need natural gas, water, and electricity lines, each of which is supplied by a different company. Can the connections be made without any crossings?

- The Five Station Quandary: Casey Jones wanted to build an elaborate model train set. He set up five stations and wanted to run tracks that connected each station directly to every other station. Could he build his layout with no crossing tracks, bridges, or shared routes?

***Exercise* 2.49.** What do the statements in Exercise **2.48** imply about the Gas-Water-Electricity Dilemma and the Five Station Quandary?

It is difficult to decide what makes a graph planar without considering non-planar graphs. We've run into two examples of non-planar graphs thus far: $K_{3,3}$, the graph describing the Gas-Water-Electric Dilemma, and K_5, the graph that represents the Five Station Quandary. If we take away any one edge from either of these graphs, we produce planar subgraphs.

***Exercise* 2.50.** Show that if we remove any one edge from either $K_{3,3}$ or K_5, the resulting subgraphs are planar.

2.6 Euler Characteristic

As we've mentioned before, sometimes hard facts can be proved by starting with simple cases and building up to more complex situations. Having control of the different ways that we can build more complex situations makes this technique even more powerful.

***Constructing Connected Graphs Algorithm* 2.51.** Every connected graph can be created by starting with a single vertex and repeatedly adding one additional edge at a time to create increasingly larger connected subgraphs until the whole graph is created.

In fact, we can improve the above theorem by specifying something about the order in which we add edges. First notice that the theorem specifies that the subgraphs stay connected at each stage, so when we add an edge e, it is of one of two types: (1) exactly one of the two endpoints of e is already in the previous subgraph or (2) both endpoints of e are already in the previous subgraph.

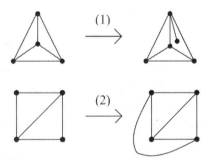

***Scholium* 2.52.** Let G be a connected graph. Then G can be constructed according to the previous theorem where all type (1) edges are added before any type (2) edge is added. If in addition G is planar, this construction allows us to redraw G using these techniques such that each partial drawing is also planar.

A graph drawn in the plane chops $\mathbb{R}^2$ into a number of regions. We will call these regions **faces**, and we will include the region "outside" the graph, called the unbounded region, as one of the faces. Each face is bounded by edges of the graph, which we call the *sides* of that face. Notice that the face inside a loop has only one side if no other part of the graph is drawn inside the loop. And the face between a pair of multiple edges has only two sides if no other part of the graph is drawn between the two edges. Even weirder, the simplest graph, which has one vertex and no edges, has one face with *no* sides.

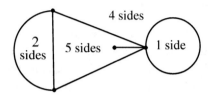

For a graph $G = (V, E)$, let $|V|$ denote the number of vertices of G and $|E|$ denote the number of edges of G. It turns out that every drawing of a planar graph in the plane will have the same number of faces as any other drawing has. So if G is planar with a fixed drawing in the plane, let $|F|$ denote the number of faces in that drawing of G. The fact that $|F|$ does not depend on the drawing of G is quite surprising from the perspective of our definition of a graph, and we will prove it shortly. For now, let's just check this assertion with some examples.

***Exercise* 2.53.** Draw a planar graph with at least five vertices and five faces. Now produce another planar drawing of the same graph that is as different as you can make it. Compare the number of faces in each drawing.

When we begin to draw a planar graph in the Constructing Connected Graphs Theorem, we start with a single vertex, no edges, and one face. As we add edges, using the two procedures in the Constructing Connected Graphs Theorem, we produce graphs that have different numbers of vertices, edges, and faces. By investigating how these two procedures change $|V|$, $|E|$, and $|F|$, we are able to say something about how these numbers are related.

***Exercise* 2.54.** Draw a graph using the two procedures detailed in the Constructing Connected Graphs Theorem. Create a chart that includes the number of vertices, number of edges, and number of faces at each stage. Do you notice any patterns?

2.6. Euler Characteristic

If you were successful with the preceding exercise, you will have discovered one of the most famous formulas in graph theory.

***Euler Characteristic Theorem* 2.55.** For a connected graph G drawn in the plane,
$$|V| - |E| + |F| = 2.$$

When a graph may not be connected, each connected piece of the graph is called a **component**.

***Corollary* 2.56.** For a graph G drawn in the plane with n components,
$$|V| - |E| + |F| = n + 1.$$

The Euler Characteristic Theorem allows us to deduce the result about the invariance of the number of faces. Notice that this next corollary does not require that the planar graph G be connected.

***Corollary* 2.57.** Let G be a planar graph. Then any two drawings of G in the plane have the same number of faces.

The Euler Characteristic Theorem also gives us a new proof of an old fact.

***Corollary* 2.58.** If G is a tree with n vertices, then G has $n-1$ edges.

The Euler Characteristic Theorem has many consequences including some theorems about the relationship between the numbers of vertices and edges of a connected planar graph.

***Theorem* 2.59.** Let G be a connected, planar graph with no loops or multiple edges having $|V|$ vertices and $|E|$ edges. If $|V| \geq 3$, then $|E| \leq 3|V| - 6$.

If the proof of the theorem above seems elusive, try this lemma first.

***Lemma* 2.60.** Let G be a planar graph with no loops or multiple edges containing at least three vertices. Then additional edges can be added to G to create a graph H where H is planar, has the same vertices as G, still has no loops or multiple edges, and all faces of H have three sides.

In general, when we consider a graph it may be difficult to prove for certain that it is impossible to draw it in the plane. How do we know that we simply haven't thought of some clever way to draw it? Conditions like those in the previous theorem on the relationship between the numbers of vertices and edges in a connected planar graph can be used to show us that certain graphs are not planar.

***Corollary* 2.61.** The graph K_5 is not planar.

The vertices in K_5 all have degree 4, so we might guess that graphs with high degree vertices are not planar. The following exercise dashes our hopes.

***Exercise* 2.62.** Construct a planar graph with no loops or multiple edges that contains no vertex of degree less than 5.

Although not all planar graphs have vertices of degree less than 5, they do have vertices of degree less than or equal to 5.

***Theorem* 2.63.** Any planar graph with no loops or multiple edges has a vertex of degree at most 5.

***Exercise* 2.64.** Theorem **2.63** requires all of its hypotheses, of which there are three. For each hypothesis, find a counterexample to the theorem if that hypothesis were removed.

We proved that K_5 is not planar. A little more analysis is required to prove that $K_{3,3}$ is not planar. Consider proving a stronger version of Theorem **2.59**.

***Theorem* 2.65.** The graph $K_{3,3}$ is not planar.

Clearly, if we have a graph built from K_5 or $K_{3,3}$ by adding vertices and edges, it cannot become planar because if we could draw the bigger graph in the plane, then that would put K_5 or $K_{3,3}$ in the plane. Also, adding extra degree 2 vertices in the middle of edges does not affect the planarity of a graph. This observation leads to the following definition.

***Definition* 2.16.** A graph $G' = (V', E')$ is a **subdivision of a graph** G if G' is obtained from $G = (V, E)$ by adding a new vertex u to V and replacing an edge $\{v, w\}$ with two edges $\{v, u\}$ and $\{u, w\}$ and repeating this process a finite number of times. Graphically, a

2.6. Euler Characteristic

subdivision G' of G is simply built by inserting zero or more vertices of degree 2 into the interiors of edges of G.

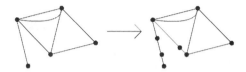

The following theorem completely characterizes whether a graph is planar or not. It turns out that planarity of graphs hinges entirely on the specific graphs $K_{3,3}$ and K_5, the graphs we know as the Gas-Water-Electricity Dilemma graph and the Five Station Quandary graph. Unfortunately, the proof of following theorem is beyond the scope of this book.

Kuratowski's Theorem. A graph G is planar if and only if G contains no subgraph that is a subdivision of $K_{3,3}$ or K_5.

***Exercise* 2.66.** For each of the following graphs, find a subgraph that is a subdivision of $K_{3,3}$ or K_5 or find a way to draw it that demonstrates that it is planar.

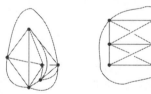

When we are thinking about actually drawing a graph in the plane, it would be delightful if each edge of the graph could be drawn as a straight segment. One interesting feature of graphs is that if they can be drawn in the plane at all, they can be drawn there with straight edges.

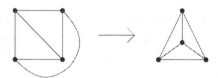

Theorem* **2.67.** Let G be a planar graph with no loops or multiple edges. Then G can be drawn in the plane in such a way that every edge is straight.

2.7 Symmetries

One of the aesthetic motivations for much wonderful mathematics is the exploration of symmetry.

Let's explore planar graphs that display pleasing symmetries. A connected, planar graph G is said to have a **symmetric planar drawing** if it has a planar drawing for which all of its vertices have the same degree and each face is bounded by the same number of edges.

***Exercise* 2.68.** Consider the three graphs below and determine which ones have a symmetric planar drawing and which do not. Justify your answers.

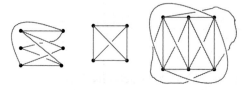

We say a graph is a **regular planar graph** if it has a symmetric planar drawing where each vertex has degree at least 3 and each face has at least 3 sides. Notice, for example, that the center graph in the previous problem is an example of a regular planar graph.

The next exercise shows that there are only a handful of regular planar graphs. Mathematicians love it when we require an object to have some property like symmetry and we're led to a finite list of possibilities.

2.7. Symmetries

***Exercise* 2.69.** Find all 5 regular planar graphs and prove that your collection is complete. (Hint: Use the Euler Characteristic.)

The previous exercise allows us to prove one of the central facts about symmetrical solids called the regular solids. A regular solid (also called a Platonic solid) is a convex, solid object with flat polygonal faces such that every face has the same number of edges and every vertex has the same degree.

***Platonic Solids Theorem* 2.70.** There are only five regular solids.

This famous result about Platonic Solids relies subtly on our ability to tell when two graphs are "the same." One of the basic notions of understanding the world and understanding mathematics arises when we ask under what circumstances two things should be considered the same. We have an intuitive idea that, if one person draws the Five Station Quandary graph and labels the vertices A, B, C, D, and E and someone else draws the Five Station Quandary graph and labels the vertices X, Y, Z, U, and W, the two graphs are really "the same." But what exactly does "the same" mean? Basically, we have just done some relabeling. So let's pin down the idea of equality of graphs.

***Definition* 2.17.** 1. Let $G = (V, E)$ and $G' = (V', E')$ be graphs without multiple edges and let $\phi_V : V \to V'$ be a function such that for every edge $\{v, w\}$ in E, $\{\phi_V(v), \phi_V(w)\}$ is an edge in E'. Then $\phi_V : V \to V'$ gives rise to a function $\phi_E : E \to E'$ naturally defined in the following way. If $\phi_V(v) = v'$ and $\phi_V(w) = w'$, then $\phi_E(\{v, w\}) = \{v', w'\}$ in E'. Putting ϕ_V and ϕ_E together gives us a function $\phi : G = (V, E) \to G' = (V', E')$ that is called a **morphism** of graphs.

2. If $\phi : G \to G'$ is a morphism of graphs where ϕ_V is a bijection between V and V' and ϕ_E is a bijection between E and E', then we say ϕ is an **isomorphism** between graphs G and G'.

3. If ϕ is an isomorphism of a graph G to itself, we call ϕ an **automorphism** of G.

These definitions allow us to be specific about what we mean when we say that two graphs are the same. A graph $G = (V, E)$ should be the same as $G' = (V', E')$ if V' is just a relabeling of

V and E' is the corresponding relabeling of E. This correspondence is exactly what the definition of an isomorphism of graphs captures. While dealing with this whole chapter, you have had an intuitive understanding of when two graphs are the same, and our definition of isomorphism has simply pinned it down. A graph automorphism captures the idea of a symmetry of the graph, since the automorphism takes the graph to itself in a one-to-one manner.

The next exercises will help you to process these definitions.

***Exercise* 2.71.** Draw two graphs that look different, but are isomorphic.

***Exercise* 2.72.** How many automorphisms does K_5 have? How many automorphisms does $K_{3,3}$ have?

***Exercise* 2.73.** For each of the 5 regular planar graphs, find all automorphisms of the graph.

If two graphs are given to you in the form of a list of vertices and a list of pairs of vertices, it is not necessarily easy to determine whether the two graphs are isomorphic.

***Exercise* 2.74.** Devise a computer program that determines whether two graphs are isomorphic.

Answering a related question would make you rich. The Subgraph Isomorphism Problem asks whether or not you can devise an *efficient* algorithm that tells, given two graphs G and H, whether H is isomorphic to some subgraph of G. If you could devise such an algorithm or could prove that no such algorithm exists, then you would have solved a famous unsolved problem called the $P = NP$ Problem, which comes with a million dollar prize! Of course, your solution would have to refer to the technical definition of efficient.

2.8 Colorability

One of the reasons to abstract a problem is that the techniques and concepts that we create when solving one problem may help us in other situations. Graph theory captures connectivity and adjacency, so questions that use these terms might benefit from graph-theoretic insights.

2.8. Colorability

lots of stuff to motivate graph coloring better, e.g. scheduling problems

For example, have you ever wondered how map makers select the colors for the countries or states on a map or globe? Well, one requirement is that adjacent countries have different colors. Under the constraint that adjacent countries have different colors, how many colors are necessary to color a map? It is the use of the word "adjacent" here that makes us think that graph theory might be useful in attacking this question.

First we need to abstract the problem and find a graph somewhere. Let's work with the continental United States for now.

There are at least two natural ways to associate a graph with a map. The first is to just make the state borders into the edges and the intersections of multiple state borders (like the Four Corners point) into the vertices. Then the coloring problem has something to do with coloring the bounded regions, the faces. This association of the map coloring problem with a graph is okay; however, this formulation is vaguely unsatisfying in that it puts the faces on center stage, whereas edges and vertices are the central ideas in graph theory. So let's find an alternative graphical representation of the map coloring problem.

When describing the problem, we said that adjacent states needed to be different colors. Recall that we used the word "adjacent" once before in this chapter, namely to describe the relationship between the endpoints of an edge. This use of the word "adjacent" suggests that we represent the states as the vertices in a graph and adjacency by edges. That is, given a map, we can put a vertex in each state and connect bordering states by an edge. That procedure gives us an alternative graph that is associated with our map.

Notice that both the graph that has the state borders as edges and the graph whose edges connect vertices inside bordering states are planar. These graphs look quite different. But they both contain the information about which states are next to which others.

***Exercise* 2.75.** Find a map of (a portion of) the United States that you can draw on, something pretty large. Construct graphs from this map using the two different procedures detailed above. Use different colored pens for the two graphs so that the two graphs are easy to distinguish visually. Describe how the two graphs are related.

These two graphs in the plane are called **dual graphs**, or more precisely, each is the dual graph of the other. Given a graph in the plane, we can draw its dual graph, as described below.

***Definition* 2.18.** Let $G = (V_G, E_G)$ be a connected planar graph with a fixed planar drawing. Construct a new graph $\hat{G} = (V_{\hat{G}}, E_{\hat{G}})$ as follows: For each face A in the drawing of G, including the unbounded one, draw a dot to represent a vertex v_A in $V_{\hat{G}}$. So the number of vertices of $\hat{G}$ equals the number of faces of G. Notice that each edge e in G has a face on each side of it in the drawing of G, say face A and face B. For each edge e in G, draw an edge that crosses e and connects v_A and v_B that represents the edge $\{v_A, v_B\}$ in $E_{\hat{G}}$. So notice that the number of edges in G equals the number of edges in $\hat{G}$. We will call the graph $\hat{G}$ the **dual of** G. Note that the edges of $\hat{G}$ can be drawn in the same plane as the drawing of G without having any of the edges of $\hat{G}$ crossing each other, so $\hat{G}$ will also be planar. We have seen that $|V_{\hat{G}}| = |F_G|, |E_{\hat{G}}| = |E_G|$, and $|F_{\hat{G}}| = |V_G|$.

***Exercise* 2.76.** Draw dual graphs for each of the following graphs. Then construct the dual graph for each of the graphs you've constructed. Do you notice anything interesting at either step?

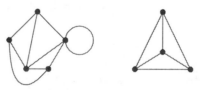

The fact that our two procedures for drawing graphs from maps produce dual graphs means that each one can be used to produce the

2.8. Colorability

other. So any information that we can glean from one can be gleaned from the other. In other words, studying one is fundamentally the same as studying the other.

Hopefully you're convinced that either procedure will do as a starting point for our abstraction from a map and that any fact we can prove about one tells a corresponding fact about its dual. The problem of coloring maps is usually considered by taking a vertex for each state and an edge between any two vertices in states that share a border. States that just touch at one point do not share a border. Any number of states could come together at one point. If we return to the map coloring problem, we must translate our challenge to refer to this new graph that we have created. In this representation, the coloring problem asks us to assign a color to each vertex such that adjacent vertices are different colors.

***Definition* 2.19.** An *n*-**coloring** of a graph is a fixed assignment of a color to each vertex such that adjacent vertices are not the same color and at most *n* colors are used. A graph is *n*-**colorable** if it has an *n*-coloring.

Notice that the definition of *n*-coloring can refer to graphs that are not planar.

***Exercise* 2.77.** Is the following graph 6-colorable? What is the smallest *n* such that this graph is *n*-colorable?

Notice that the map coloring problem is simple for maps with a small number of states. Certainly, if we wanted to color a map with only 6 colors and if there were only 6 or fewer states on our map, it would be easy.

***Theorem* 2.78.** For any natural number, n, let $G = (V, E)$ be a graph with $|V| \leq n$ that has no loops. Then G is n-colorable.

Similarly, if you have more colors than a vertex could be adjacent to, coloring is easy.

Theorem 2.79. Let $G = (V, E)$ be a graph with no loops. Let $M = \max\{\deg(v) | v \in V\}$, which is the biggest degree of a vertex. Then G is $(M + 1)$-colorable.

The remainder of the section is dedicated to proving that 5 colors are enough to color *any* planar graph, that is, 5 colors are sufficient to color the vertices of any planar graph such that no two adjacent vertices have the same color. Our strategy for proving this theorem is to isolate conditions under which it is possible to extend a 5-coloring of a subgraph to a 5-coloring of a larger graph.

Theorem 2.80. Consider a graph, G, that is built from a subgraph, H, by adding one new vertex, v, and new edges that connect the new vertex to vertices in H. If the subgraph H has a 5-coloring such that the new vertex, v, is not adjacent to vertices of all five colors, then G is 5-colorable.

One circumstance under which a new vertex will not be adjacent to vertices of all five colors is when the new vertex is not adjacent to five vertices altogether.

Lemma 2.81. If a graph G has no loops and G is the union of a 5-colorable subgraph, H, and a new vertex, v, with its edges such that v has $\deg_G(v) < 5$, then G is 5-colorable.

The combination of this last theorem and lemma suggests the following inductive approach to answering the map coloring problem. We will color the vertices one at a time, being careful that the next vertex is only adjacent to vertices with at most 4 colors. But sadly, not all planar graphs have a vertex of degree less than or equal to 4, so this procedure may not always be possible. However, this approach is sufficient to prove the Six Color Theorem. For this theorem, you need to know that certain planar graphs have vertices with degree less than or equal to 5, which we proved in Theorem **2.63**.

Six Color Theorem 2.82. Any planar graph with no loops is 6-colorable.

2.8. Colorability

In what remains, we must deal with the situation in which we are forced to color a degree 5 vertex. We will need some notation to talk about the portions of a graph that use certain colors.

***Definition* 2.20.** Let H be a 5-colorable graph with a fixed coloring, and let S be a subset of the colors. Then define H_S to be the subgraph of H that contains all of the vertices with colors in S and all of the edges both of whose endpoints are vertices with colors in S.

***Exercise* 2.83.** For the graph H below, which has been colored with the colors $\{r, b, y, g\}$, construct $H_{\{r,y\}}$, $H_{\{r,b,g\}}$, and $H_{\{y\}}$.

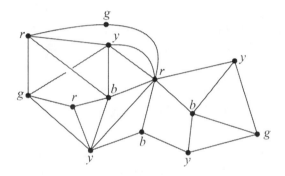

A graph might have several different n-colorings, and selecting a good one could be useful.

***Exercise* 2.84.** Give an example of a graph with two very different 3-colorings.

Let's consider how we can change a fixed coloring to produce a new one with desired properties. In the lemma below, you will have to change the 5-coloring of H in order to be able to color the vertex v.

***Lemma* 2.85.** Let G be a graph without loops that is the union of a 5-colorable subgraph, H, and one additional vertex v of degree 5 and five edges $\{v, v_1\}, \{v, v_2\}, \{v, v_3\}, \{v, v_4\},$ and $\{v, v_5\}$ where each vertex v_i is in H. Fix a 5-coloring of H such the vertices v_1, v_2, v_3, v_4, v_5 are colored c_1, c_2, c_3, c_4, c_5 respectively. Suppose that v_i and v_j are not connected in $H_{\{c_i,c_j\}}$ for some pair of vertices/ colors. Then G is 5-colorable.

Proving the preceding lemma involves writing down a procedure for finding a 5-coloring of G given the hypotheses. Is your procedure written so that a person or computer could use it to actually find the 5-coloring of G? If not, you're not done yet.

When G is planar, the situation described in the previous lemma must occur.

Lemma 2.86. Let G be a planar graph without loops or multiple edges that is the union of a 5-colorable graph H and one additional vertex v of degree 5 and five edges $\{v, v_1\}, \{v, v_2\}, \{v, v_3\}, \{v, v_4\}$, and $\{v, v_5\}$ where each vertex v_i is in H. Fix a 5-coloring of H and suppose the five adjacent vertices to v, v_1, v_2, v_3, v_4, v_5, are in cyclic order around v and have colors c_1, c_2, c_3, c_4, c_5 respectively. Then either v_1 and v_3 are not connected in $H_{\{c_1, c_3\}}$, or v_2 and v_4 are not connected in $H_{\{c_2, c_4\}}$.

Use the lemmas above to prove the following 5-color theorem.

Five Color Theorem 2.87. Any planar graph with no loops is 5-colorable.

Corollary 2.88. Any map with connected countries can be colored with five colors such that no two countries that share a border have the same color.

In fact, four colors suffice to color any map. The following Four Color Theorem was a famous unsolved problem for more than a hundred years before it was proved using exhaustive computer methods. Its proof uses the Euler Characteristic Theorem extensively as well as techniques like those that we developed for switching colorings in H above, but the proof is extremely complicated and some consider it a bit unsatisfying in that the proof involves many cases that can be checked only by computers.

Four Color Theorem. Any planar graph with no loops is 4-colorable.

2.9 Completing the Walk around Graph Theory

In this unit we explored the Königsberg Bridge Problem, the Paperperson's Puzzle, the Gas-Water-Electricity Dilemma, the Five Station Quandary, and the Map Coloring Challenge. We defined a graph, a

2.9. Completing the Walk around Graph Theory

walk, and a circuit to help us investigate traceability. We talked about planarity of a graph and found the relationship among the numbers of vertices, edges, and faces that was captured in the Euler Characteristic. These investigations let us deduce results not only about graphs in the plane, but also about the five Platonic solids and colorability of maps.

Graph theory has many other applications, ranging from computer circuitry to epidemics to social networking. Our introduction to graph theory has merely cracked open a door behind which you can find many further delights.

In addition to producing interesting results about graphs, this unit illustrates how the strategy of mathematical inquiry can create rich, new ideas. Let's reflect on the journey we've taken and lessons we've learned by highlighting the critical strategies we used along the way.

- **Abstraction** – Our investigation began with the challenge of the Königsberg Bridge Problem. We searched for other challenges that have similar core properties, and we found the Paperperson's Puzzle, the Gas-Water-Electricity Dilemma, the Five Station Quandary, the Map Coloring Challenge, and others (**2.1, 2.6**). We distilled this core similarity into the definition of a graph (Definition **2.1**) and translated the challenges into precise questions (**2.16**, Definition **2.6**, Definition **2.8, 2.30**).

    motivated by problem

- **Exploration of Definition** – We explored the definition of a graph by checking that our motivating examples do produce graphs (**2.2, 2.3, 2.8**), generating new trivial and non-trivial examples (**2.9, 2.13**), clarifying special or degenerate cases of the definition (Definition **2.2**, Definition **2.3**), establishing appropriate notation and vocabulary for working with the definition (**2.2, 2.4, 2.5**, Definition **2.1**, Definition **2.2**, Definition **2.3**, Definition **2.4**, Definition **2.5**, Definition **2.6**, Definition **2.7**), and stating and proving basic theorems about the definition (**2.10, 2.11**).

- **Investigation, Justification, and Application** – We set out to solve our initial challenge by generalizing the precise question (**2.15, 2.18**), systematically generating examples and using them to explore the generalized challenge, conjecturing answers to the generalized challenge, giving an intuitive reason for believing the conjectures (**2.19**, **2.20**, **2.21**), transforming those intuitive reasons into rigorous proofs (**2.31**, **2.35**), verifying those rigorous proofs in special cases including the case of the original precise challenge (**2.32**, **2.37**), and applying that work to solve the original challenge (**2.33**, **2.34**).

- **Extension** – We continued our exploration by identifying new areas revealed by this inquiry and repeating this process in those domains.

With every development of ideas that arise from a different set of motivating questions we add more tools to our toolkit of mathematical inquiry.

3

Groups

3.1 Examples Lead to Concepts

One of the most powerful and effective methods for creating new ideas is to look at familiar parts of our world and isolate essential ingredients. In mathematics, this strategy is particularly effective when we find several familiar examples that seem to share common features. So we will begin our next exploration by looking carefully at *adding*, at *multiplying*, and at *moving blocks*, with an eye toward finding similarities.

(1) Adding: Among the first computational skills we learn in our youths is addition of integers. So our first example is the familiar integers accompanied, as they are, by the method of combining them through addition.

(2) Multiplying: Real numbers are among our next mathematical objects and multiplication is a method of combining a pair of numbers to produce another number.

(3) Moving blocks: This example involves an equilateral triangular block fitting into a triangular hole, presenting challenges that you might recall from the first years of your life. As an inquisitive toddler, you explored all the different ways of removing the block from the hole and replacing it. You could just put it back in the same position. You could rotate it counterclockwise by 120 degrees and put it back

44 **3. Groups**

in the hole. You could rotate it counterclockwise by 240 degrees and put it back. You could flip it over leaving the top corner fixed. You could flip it over leaving the bottom left corner fixed. Or you could flip it over leaving the bottom right corner fixed. You could combine two motions of the block by first doing one and then doing the second, that is, you could compose two transformations to form another transformation.

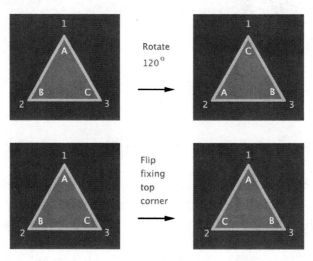

Now let's undertake a mathematical exploration by seeking the essential and isolating common features of these examples. All three examples involve combining two elements to get a third. In the case of addition of integers, we add two integers to get another integer; e.g., $2+3 \mapsto 5$. In the case of multiplication of reals, we multiply two reals to get another real number; e.g., $3 \cdot 1.204 \mapsto 3.612$. In the case of ways to move the block, we combine two transformations of the block to get a third; e.g., [flip it over leaving the top corner fixed] ∘ [flip it over leaving the bottom right corner fixed] $\mapsto$ [rotate it counterclockwise by 240°]. Traditionally, if S and T are two transformations, then $T \circ S$ means to perform the transformation S *then* the transformation T. If you read the symbol "∘" as "after", you will do the transformations in the correct order. Be careful to respect this convention.

What all of these examples have in common is that in each case we start with some collection (integers, reals, transformations of a tri-

3.1. Examples Lead to Concepts

angle) and we have some operation (addition, multiplication, composition) that takes any two items from the collection and returns a third. Because our operations take *two* elements as input, we call them *binary* operations. Our rules for combining have some other features in common as well.

First common feature—an identity element: In each of our examples there is an element that, when combined with any other element, has no effect on the other element. We call that "ineffective" element an *identity element*.

(1) In addition of integers: $0 + 3 = 3$. In fact, $0 + n = n = n + 0$ for every integer n.

(2) In multiplication of reals: $1 \cdot 2.35 = 2.35$. In fact, $1 \cdot r = r = r \cdot 1$ for every real number r.

(3) In composing transformations of a triangular block: [just put the block back unchanged] ∘ [rotate counterclockwise by 120°] = [rotate counterclockwise by 120°]. In fact, [just put the block back unchanged] ∘ $T = T = T$ ∘ [just put the block back unchanged] for any transformation T of the block.

Second common feature—inverses: In each example, every element can be combined with another element to produce the identity element; that is, for each element there is another that undoes it. This "reversing" element is called an *inverse*.

Actually, in our example of the reals not every element has an inverse because nothing times 0 gives 1. So we will change the example of the reals under multiplication a little, namely, we will omit 0. The set in our second example will now be all the real numbers except 0. This process of modifying our examples in the face of difficulties has led to lots of interesting mathematics.

(1) In addition of integers: $3 + (-3) = 0$. In fact, for every integer n,

$$n + (-n) = 0 = (-n) + n.$$

(2) In multiplication of reals except 0: $2.35 \cdot \frac{1}{2.35} = 1$. In fact, for every non-0 real number r,

$$r \cdot \frac{1}{r} = 1 = \frac{1}{r} \cdot r.$$

(3) In composing transformations of a triangular block: [rotate counterclockwise by 120°] ∘ [rotate counterclockwise by 240°] = [just put the block back unchanged]. In fact, every transformation of the block can be followed with another transformation that returns the block to its previous position.

***Exercise* 3.1.** 1. Show that there are exactly six transformations of the equilateral triangle. To save some writing, let's use the following notation for the six transformations:

1. R_0 = [just put the block back unchanged]

2. R_{120} = [rotate counterclockwise by 120°]

3. R_{240} = [rotate counterclockwise by 240°]

4. F_T = [flip it over leaving the top corner fixed]

5. F_L = [flip it over leaving the bottom left corner fixed]

6. F_R = [flip it over leaving the bottom right corner fixed]

2. Make a chart that lists each of the six transformations of the equilateral triangle and, for each transformation, its inverse.

Third common feature—associativity: Any rule for combining a pair of elements to get a third leaves us with an intriguing ambiguity about how three elements might be combined. When adding three integers, what do we do? Stop and compute $2 + 4 + 6$; try to explain what you did. You probably replaced the one addition question by a sequence of problems you knew how to deal with: first add two of the integers and then add the result to the third. Similarly, when adding any number of integers, we actually perform a sequence of additions of two numbers at a time.

3.1. Examples Lead to Concepts

So what does the expression $k+m+n$ really mean? There are two different ways to break this expression down into a sequence of pairwise addition problems: $(k+m)+n$ or $k+(m+n)$. Parentheses mean what they always have, namely, the order of operations goes from inside the parentheses to outside. Both of these possible sequences are reasonable ways of reducing a question of adding three integers down to the case of adding pairs of integers sequentially. In our examples, the choice of sequencing doesn't matter. More precisely, in each example, both choices of ways to put parentheses on $k+m+n$, $r \cdot s \cdot t$, or $R \circ S \circ T$ produce the same result. This feature of the operation is called *associativity*.

(1) In addition of integers: for any three integers k, m, and n,

$$(k+m)+n = k+(m+n).$$

(2) In multiplication of reals except 0: for any three non-0 reals r, s, and t,

$$(r \cdot s) \cdot t = r \cdot (s \cdot t).$$

(3) In composing transformations of a triangular block: for any three transformations R, S, and T,

$$(T \circ S) \circ R = T \circ (S \circ R).$$

This last fact is not completely obvious, so you might fear that you'd have to verify it by laboriously checking every possible sequence of three transformations. Fortunately, there is an easier way. These transformations are functions, namely, each transformation is a function whose domain and range is the set {top corner, bottom left corner, bottom right corner}. For example, the R_{120} transformation can be thought of as the function that takes the top corner to the bottom left corner, the bottom left corner to the bottom right corner, and the bottom right corner to the top corner. It is straightforward to show that the composition of functions is associative as long as the composition is defined. The fact that transformations are functions also explains the order convention of $T \circ S$ as T after S.

***Exercise* 3.2.** In composing transformations, check an example of associativity by confirming the following equality:

$$(F_T \circ R_{120}) \circ F_L = F_T \circ (R_{120} \circ F_L).$$

If this equality seems completely trivial, then you are not being careful with the order of operations.

***Exercise* 3.3.** Let R, S, T, and U be transformations of a triangular block, and consider the expression $U \circ T \circ S \circ R$. How many different ways are there to put parentheses on this expression such that only two transformations are composed at a time? (Do not change the order of the transformations; only add parentheses to the expression.)

Let's note one feature that is not shared by all three of our examples. In the example of the integers under addition, for any integers m and n, $m + n = n + m$. Likewise, in the example of the reals under multiplication, for any real numbers r and s, $r \cdot s = s \cdot r$; however, notice that the order makes a difference in composing transformations of the triangle.

***Exercise* 3.4.** Find some examples of two transformations of an equilateral triangle where composing the transformations in one order gives a different result from doing them in the other order. Each of your examples should be a pair of transformations of the triangle, S and T, such that $S \circ T \neq T \circ S$.

When the order does not matter, that is, when we always get the same result no matter in which order we do the binary operation, then we call the operation **commutative**. We will talk more about this distinction later, but from Exercise **3.4** we know that it is possible that the same two elements combined in the opposite order might yield a different result.

Now let's take a step that creates mathematical ideas, namely, defining a concept that captures the common features that we have found. It turns out that we have isolated the essential ingredients of a mathematical structure that is called a *group*. We'll give the definition here and then make sure that we have pinned down all the features thoroughly.

3.1. Examples Lead to Concepts

***Definition* 3.1.** A **group** is a set G with a binary operation $*$, written $(G, *)$, such that:

1. The operation $*$ is closed and well-defined on G.

2. The operation $*$ is associative on G.

3. There is an element $e \in G$ such that $g * e = g = e * g$ for all $g \in G$. The element e is called the **identity**. In particular, G is non-empty.

4. For each element $g \in G$ there is an element $h \in G$ such that $g * h = e = h * g$. This element h is called the **inverse of** g and is often written as g^{-1}.

Our examples have given us an intuitive idea of what we want to convey, but we may want to take a further step of precision. Since the terms "binary operation," "closed," and "well-defined" may not be completely clear yet, we will describe them a bit more and then have an exercise that helps to elucidate them.

A binary operation is a procedure that takes two elements from a set and returns a third object. It is possible that this third object does not lie in our original set; if this happens, we say that the binary operation is not *closed*. Also, if there is some choice in how we refer to the elements of the set (such as having different ways of referring to the same rational number) and if the binary operation is given in terms of a rule that is dependent on how we refer to the elements of the set, then the rule might return different values even though the input has not changed. If the operation suffers from this kind of ambiguity, we say that the operation is not *well-defined*.

Here is an exercise to help you clarify the ideas of binary operation, closed, and well-defined.

***Exercise* 3.5.** Show that the following operations ($*$) are or are not closed, well-defined binary operations on the given sets. And remember to justify your work: exercises are just specific theorems.

1. The interval $[0, 1]$ with $a * b = \min\{a, b\}$

2. $\mathbb{R}$ with $a * b = a/b$

3. $\mathbb{Z}$ with $a * b = a^2 + b^2$

4. $\mathbb{Q}$ with $a * b = \dfrac{\text{numerator of } a}{\text{denominator of } b}$

5. $\mathbb{N}$ with $a * b = a - b$

Remember that a group $(G, *)$ is a set together with a closed, well-defined, associative binary operation with an identity element, e, and, for each $g \in G$, an inverse element g^{-1}.

Let's begin our exploration of this new mathematical entity, a group, by first recording that our generative examples are groups. There is no need to prove these theorems now.

Theorem. The integers with addition, $(\mathbb{Z}, +)$, is a group.

Theorem. The nonzero real numbers with multiplication, $(\mathbb{R} \setminus \{0\}, \cdot)$ is a group.

Theorem. The transformations of an equilateral triangle in the plane with composition is a group. We call this group $(D_3, \circ)$, the symmetries of the equilateral triangle.

When we write theorems about an arbitrary group $(G, *)$, we will often write G for $(G, *)$ to simplify the notation; we know that G has a binary operation, but we don't explicitly name it. Similarly, we will sometimes write gh when we mean $g * h$.

Now that we have the definition of a group, we are interested in describing the full spectrum of possibilities that it represents. In other words, we want to classify all groups. There are many directions from which we could approach this goal; we start by proving theorems about *all* groups in order to notice properties that must be common to everything in the spectrum. Our first theorem that is true for any group tells us that a group can have only one identity element.

Theorem 3.6. Let G be a group. There is a *unique* identity element in G. In other words, there is only one element in G, e, such that $g * e = g = e * g$ for all g in G.

Every group satisfies the following Cancellation Law. It seems simple and obvious, but it is an extremely useful property; it will reappear in the proof of every important theorem for the duration of this chapter.

3.1. Examples Lead to Concepts

***Cancellation Law* 3.7.** Let G be a group, and let $a, x, y \in G$. Then $a * x = a * y$ if and only if $x = y$. Similarly, $x * a = y * a$ if and only if $x = y$.

Be careful not to use the theorem when proving it. Instead, only use properties given to you by the definition of a group. As always, the phrase "if and only if" means that there are actually two theorems involved. To prove $a * x = a * y$ if and only if $x = y$, you need to prove that $a * x = a * y$ implies $x = y$, *and* you need to prove $x = y$ implies $a * x = a * y$.

***Exercise* 3.8.** Show that the Cancellation Law fails for $(\mathbb{R}, \cdot)$, thus confirming that $(\mathbb{R}, \cdot)$ is not a group.

***Corollary* 3.9.** Let G be a group. Then each element g in G has a *unique* inverse in G. In other words, for a fixed g, there is only one element, h, such that $g * h = e = h * g$.

Recall that in general $g * h$ may not equal $h * g$; however, if one product is the identity, then both orders of the product yield the identity.

***Theorem* 3.10.** Let G be a group with elements g and h. If $g * h = e$, then $h * g = e$.

In words, this theorem says that in a group, if h is a right inverse of g, then it is also a left inverse of g. So we only need to check that g and h are one-sided inverses to know that they are inverses.

***Theorem* 3.11.** Let G be a group and $g \in G$. Then $(g^{-1})^{-1} = g$.

Theorems like the preceding four show us that if we have a structure that satisfies the definition of a group, then it will automatically have the features stated in the theorems. One of the strategies and strengths of abstract mathematics is that we define a structure (like a group) and then deduce that any mathematical object of that type (any group, for example) will have features (like a unique identity or the cancellation property) that are common to every example of such a structure (every group).

3.2 Clock-Inspired Groups

In order to develop our intuition about groups, let's first consider a few more examples that we can create by taking our existing examples and seeking variations. Taking examples and concepts that we have and making variations of them is one of the most common and most powerful methods for creating new mathematical ideas. In this section, we make two families of examples related to adding integers and telling time.

Our first example of a group was the integers with the binary operation of addition, $(\mathbb{Z}, +)$. In life we also perform addition of numbers when we tell time, but in that case we have a cyclical kind of addition. If it's 9 o'clock now, then in 47 hours it will be 8 o'clock. Somehow in our world of time, "$9 + 47 = 8$". Can we construct a group that captures this cyclical kind of arithmetic? Well, we know what we need to construct a group: we need a set of elements and a binary operation. So to construct a group that captures the idea of times of the day, we might consider the hours $\{1, 2, 3, 4, 5, 6, 7, 8, 9, 10, 11, 12\}$ as our set and *clock addition* as our operation. Notice that this definition of clock addition only allows us to combine two numbers from 1 to 12, so we could not add 47 to a time, for example. We'll deal with that issue later. For now, we have created a group.

***Exercise* 3.12.** If $G = \{1, 2, 3, 4, 5, 6, 7, 8, 9, 10, 11, 12\}$ and $\oplus_{clock}$ is the binary operation clock addition, show that $(G, \oplus_{clock})$ is a group. What is the identity element? What is the inverse of 3?

Once we have defined this clock group, we cannot (and should not) resist the urge to extend the idea. An obvious and important way to generalize the idea is to consider clock arithmetic with different numbers of hours in the day. That generalization gives us infinitely many different groups that use cyclical addition. Let's now take the step of pinning down all these ideas with formal definitions.

Let $C_n = \{0, 1, \ldots, n-1\}$. We will define a binary operation on C_n that captures the idea of cyclical arithmetic. Any two elements in C_n are integers, so we can add them. If their sum is strictly less than n, then it is in C_n, so their sum makes sense as an element of C_n. If their sum is m, bigger than or equal to n, then replace it by $m - n$,

3.2. Clock-Inspired Groups

which is now definitely back in the set. Call this operation *n-cyclic addition*, and write it is as $\oplus_n$. In other words, if a and b are elements of C_n, then

$$a \oplus_n b = \begin{cases} a+b & \text{if } a+b < n \\ a+b-n & \text{if } n \leq a+b \end{cases}.$$

***Theorem* 3.13.** For every natural number n, the set C_n with n-cyclic addition, $(C_n, \oplus_n)$, is a group. We call it the **cyclic group of order** n.

Our cyclic groups are nice, but somehow we need to deal with the fact that in reality we can add 47 hours to a time. How can we extend the description of our clock world so as to include 47 and other integers in it? A solution is presented to us by considering European and military time, where time is measured with a 24 hour clock rather than a 12 hour clock. When their clocks read 15 o'clock, ours read 3 o'clock. This idea of reducing by 12 can easily be extended even to 47. What time is 47 o'clock? Answer: it's 11, because $47 - 12$ is 35, $35 - 12$ is 23, and $23 - 12$ is 11. More simply put, since $47 = 3 \cdot 12 + 11$ or, equivalently, $47 - 11 = 3 \cdot 12$, we consider 47 and 11 to be referring to the same time, that is to say, 47 and 11 should be different names for the same element of our group that captures the idea of time. In general, we could say that two integers a and b are equivalent if $a = b + 12k$ for some integer k or, saying the same thing, $a - b$ is a multiple of 12. With this idea in mind, we can think of a new group with twelve elements $\{[1], [2], [3], \ldots, [12]\}$; however, each element really stands for all the integers that are equivalent to it using our concept of time equality.

***Definition* 3.2.** Let n be a natural number. Two integers a and b are said to be **congruent modulo** n if there exists an integer k such that $a = b + kn$ or, equivalently, $a - b = kn$. In other words, two integers are congruent modulo n if and only if their difference is divisible by n. We write "a is congruent to b modulo n" as $a \equiv b \mod n$.

We can now define a set $\mathbb{Z}_n$ that contains n elements, but each of those n elements can be referred to in infinitely many different ways.

***Definition* 3.3.** Let

$$\mathbb{Z}_n = \{[a]_n \mid a \in \mathbb{Z}, [a]_n = [b]_n \text{ if and only if } a \equiv b \mod n\}.$$

Then define the binary operation $\oplus$ on $\mathbb{Z}_n$ by $[a]_n \oplus [b]_n = [a+b]_n$, which we will call **modular addition**.

Now we can try answering the question, "What time is it 47 hours after 9 o'clock?" Using modular arithmetic, we can replace the question with $[9]_{12} \oplus [47]_{12} = [56]_{12} = [8 + 4(12)]_{12} = [8]_{12}$.

Exercise **3.14.** Show that $\oplus$ is well-defined on $\mathbb{Z}_n$. That means, show that if you replace integers a and b by congruent integers a' and b' respectively, then $[a]_n \oplus [b]_n = [a']_n \oplus [b']_n$.

Note that both $(C_n, \oplus_n)$ and $(\mathbb{Z}_n, \oplus)$ are groups with n elements and a cyclical addition. Intuitively, they are clearly the "same", but we do not yet have a definition for when two groups are the same. We will return to this issue later, but for now, notice that when working with C_n, the group's elements look like $\{0, 1, \ldots, n-1\}$, and the operation is called n-cyclic addition, written $\oplus_n$. When working with $\mathbb{Z}_n$, the distinct elements look like $\{[0]_n, [1]_n, \ldots, [n-1]_n\}$, and the operation is called modular addition, written $\oplus$.

Given a group, a table that lists all group elements and the results of the binary operation on any pair is called a Cayley table. For example, the following chart together with the sentence that explains how to interpret the chart is the Cayley table for $(C_5, \oplus_5)$.

$\oplus_5$	0	1	2	3	4
0	0	1	2	3	4
1	1	2	3	4	0
2	2	3	4	0	1
3	3	4	0	1	2
4	4	0	1	2	3

To find $a \oplus_5 b$ on this table, locate the row starting with a and the column starting with b and find their intersection.

Exercise **3.15.** How could you modify the above Cayley table to make it a Cayley table for $\mathbb{Z}_5$?

Exercise **3.16.** Looking at the Cayley tables for C_5 and $\mathbb{Z}_5$, do you notice any feature of the rows and columns that could be generalized to all Cayley tables? Make a conjecture and prove it.

3.3 Symmetry Groups of Regular Polygons

The symmetries of an equilateral triangle under composition form one of our generative examples of a group. We called this group D_3 because it consisted of transformations of a shape in the plane with 3 sides, having uniform side lengths and angles. Similarly, we could think about the transformations of the shape with 4 equal sides and uniform angles, more commonly known as the square. Of course, we could also consider similar shapes with any number of sides. So we can create related groups by considering the symmetries of any *regular polygon*. By a **symmetry**, we mean a transformation that takes the regular polygon to itself as a rigid object.

***Exercise* 3.17.** Show that every transformation that takes a regular polygon to itself as a rigid object is either a rotation or a reflection.

Each symmetry of a regular polygon can be viewed as a function whose domain and range is the set of vertices of the polygon, so we can use composition as the binary operation.

***Exercise* 3.18.** Consider a square in the plane. How many distinct symmetries does it have? Give each symmetry a concise, meaningful label. For every pair of symmetries, S and T, compose them in both orders, $S \circ T$ and $T \circ S$. Record all of this information in a Cayley table.

The following theorem notes that we have a new group.

***Theorem* 3.19.** The symmetries of the square in the plane with composition form a group.

The group of symmetries of the regular 4-gon (i.e., the square) with the binary operation of composition is denoted D_4. In general, the symmetries of the regular n-gon form a group, which is denoted D_n and called a **dihedral group**.

Exercise 3.20. For each natural number n, how many elements does D_n have? Justify your answer.

3.4 Subgroups, Generators, and Cyclic Groups

The last two sections have helped us find new (families of) groups by modifying and generalizing existing examples. This process has increased the variety of groups that is familiar to us tremendously. In this section, we continue to expand our repertoire of groups, but in a new direction. Here we look for new groups inside the ones we know and investigate the ways that groups can be built up from within.

Definition 3.4. A **subgroup** of a group $(G, *)$ is a non-empty subset H of G along with the restricted binary operation such that $(H, *|_H)$ is a group.

To check that a subset is a subgroup requires us to check all of the conditions of a group, though more attention is often paid to the subset being closed under $*$ and to inverses being in the subset. To check that a subset is closed under $*$, we must confirm that, for any pair of elements in the subgroup, the binary operation performed on them results in another element in the subgroup.

Let's take a look at our examples and find some of their subgroups.

Exercise 3.21. 1. Show that the even integers, written $2\mathbb{Z}$, form a subgroup of $(\mathbb{Z}, +)$. Technically, $2\mathbb{Z}$ is just a set, but we will often drop the binary operation from the notation for a subgroup when it is obvious.

2. Show that the set of nonzero rational numbers, $\mathbb{Q} \setminus \{0\}$, is a subgroup of $(\mathbb{R} \setminus \{0\}, \cdot)$.

3. Show that the set of three transformations $H = \{R_0, R_{120}, R_{240}\}$ is a subgroup of D_3, the symmetries of the triangle.

4. Show that $K = \{0, 15, 30, 45\}$ is a subgroup of $(C_{60}, \oplus_{60})$.

If a group is defined by a set of elements satisfying a certain condition, then a subgroup is usually a subset satisfying a stronger condition. In the exercise above, the *even integers* form a subgroup of *all integers*, the nonzero *rational* numbers are a subgroup of the nonzero

3.4. Subgroups, Generators, and Cyclic Groups

reals, the *rotations* are a subgroup of *all symmetries*, and the *quarter hours* are a subgroup of *all minutes* in a group capturing 60-minute clock arithmetic.

But not every condition defines a subgroup. For example, the odd integers are not a subgroup of the integers under addition. (Why not?) To get a sense of what subsets of a group form a subgroup, it is a good exercise to describe all the subgroups of a few groups.

Exercise 3.22. For each of the following groups, find all subgroups. Argue that your list is complete.

1. $(D_4, \circ)$

2. $(\mathbb{Z}, +)$

3. $(C_n, \oplus_n)$

You may have noticed that the identity element is in each of your subgroups.

Theorem 3.23. Let G be a group with identity element e. Then for every subgroup H of G, $e \in H$.

The smallest and simplest subgroup of any group is just the identity element.

Theorem 3.24. Let G be a group with identity element e. Then $\{e\}$ is a subgroup of G.

Every group is a subgroup of itself; this subgroup is necessarily the biggest subgroup.

Theorem 3.25. Let G be a group. Then G is a subgroup of G.

Any group G has the subgroups $\{e\}$ and G, so these subgroups are basically trivial. If H is a subgroup of G such that $\{e\} \subsetneq H \subsetneq G$, then we say that H is a **non-trivial** subgroup of G.

Definition 3.5. For repeated applications of the binary operation to one element g in a group G, we will sometimes use exponents: $g^4 = g*g*g*g$. In general, if n is a positive integer, then g^n is the binary operation applied to n copies of g and g^{-n} is the binary operation applied to n copies of g^{-1}, which, you recall, is our notation for the inverse of g. We define g^0 to be e. Note that $g^1 = g$.

We have some intuition about how exponents work, and that intuition is the reason for this short-hand notation, but be careful not to use any properties of exponents that you have not checked for groups.

Exercise 3.26. Let G be a group, $g \in G$, and $n, m \in \mathbb{Z}$. Then

1. $g^n g^m = g^{n+m}$, and
2. $(g^n)^{-1} = g^{-n}$.

Definition 3.6. Let G be a group and g be an element of G. Then $\langle g \rangle$ is the subset of elements of G formed by repeated applications of the binary operation using only g and g^{-1}, that is,

$$\langle g \rangle = \{g^{\pm 1} * g^{\pm 1} * \cdots * g^{\pm 1}\}.$$

Notice that using the notation of the previous exercise,

$$\langle g \rangle = \{g^m | \text{ for all } m \in \mathbb{Z}\}.$$

Theorem 3.27. Let G be a group and g be an element of G. Then $\langle g \rangle$ is a subgroup of G.

We call $\langle g \rangle$ the **subgroup of G generated by g**.

Exercise 3.28. Show that the subgroup $2\mathbb{Z}$ of $(\mathbb{Z}, +)$ is the subgroup generated by 2 (or -2); that is, show that $2\mathbb{Z} = \langle 2 \rangle = \langle -2 \rangle$.

The subgroups $\langle g \rangle$ are generated by a single element g in a group G. In a similar way, we can consider subgroups generated by more than one element. If S is any subset of a group G, then we define $\langle S \rangle$ to be all elements of G that are obtained from finite combinations of elements of S and their inverses, $\{s_1^{\pm 1} * s_2^{\pm 1} * \cdots * s_n^{\pm 1} | s_i \in S\}$.

Theorem 3.29. Let G be a group and S be a subset of G. Then $\langle S \rangle$ is a subgroup of G. Moreover, if H is a subgroup of G and $S \subset H$, then the subgroup $\langle S \rangle$ is a subgroup of H.

3.4. Subgroups, Generators, and Cyclic Groups

As was the case with a single element, $\langle S \rangle$ is called the subgroup generated by S. It is the smallest subgroup that contains all the elements of S, and $\langle g \rangle$ is the smallest subgroup that contains g.

The preceding theorems show us a method for constructing a subgroup of any group. We can just start with any collection of elements from the group and then look at all the elements we get by performing the binary operation repeatedly on those elements and their inverses.

***Exercise* 3.30.** In Exercise **3.22**, you described the subgroups of $(\mathbb{Z}, +)$. Which subgroup of $(\mathbb{Z}, +)$ is $\langle \{6, -8\} \rangle$? Which subgroup is $\langle \{5, -8\} \rangle$?

Understanding generators is very useful for our goal of classifying the wide range of groups because all groups are generated by their elements. Generators are tools that can help us understand the structure of every group. This realization prompts us to ask about those groups that have a small number of generators.

***Definition* 3.7.** A group G is called **cyclic** if there is an element g in G such that $\langle g \rangle = G$. In other words, a group is cyclic if it is generated by one element.

Some of the groups that we have considered are cyclic.

***Theorem* 3.31.** The integers under addition, $(\mathbb{Z}, +)$, is a cyclic group.

We named $(C_n, \oplus_n)$ the "cyclic group of order n", and we are now prepared to partially justify that name.

***Theorem* 3.32.** For every natural number n, the groups C_n and $\mathbb{Z}_n$ are cyclic groups.

In general, subgroups of groups can be complicated; however, each subgroup of a cyclic group is generated by one element.

***Theorem* 3.33.** Any subgroup of a cyclic group is cyclic.

Not all groups are cyclic. The groups D_n give us some examples of non-cyclic groups.

***Theorem* 3.34.** The groups D_n for $n > 2$ are not cyclic.

Although the D_n groups are not cyclic, they are generated by just two elements.

Exercise **3.35.** For each natural number n, find a pair of elements that generate D_n.

Definition **3.8.** A group G is called **finite** if the underlying set is finite. Similarly, G is called **infinite** if its underlying set is infinite. A group is **finitely generated** if $G = \langle S \rangle$ for some finite subset, S, of its elements.

Theorem **3.36.** Every finite group G is finitely generated.

Theorem **3.37.** The group $(\mathbb{Q} \setminus \{0\}, \cdot)$ is not finitely generated.

Theorem* **3.38.** The group $(\mathbb{R} \setminus \{0\}, \cdot)$ is not finitely generated.

Definition **3.9.** The number of elements in (the underlying set of) G is called the **order** of G, written $|G|$. The order of an element g, written $o(g)$, is the order of the subgroup that it generates,

$$o(g) = |\langle g \rangle|.$$

Exercise **3.39.** Compute the order of each element $T \in D_4$. Carefully use the *definition* of $o(T)$.

The order of an element g of a group G is defined in terms of the number of elements in $\langle g \rangle$; however, that number is also the smallest power of the element g that equals the identity element of the group G.

Lemma **3.40.** Let g be an element of a finite group G with identity e. Then there is a natural number m such that $g^m = e$.

Theorem **3.41.** Let g be an element of a finite group G whose identity element is e. Then $o(g)$, which is defined as $|\langle g \rangle|$, is the smallest natural number r such that $g^r = e$.

One fundamental difference between the structures of $(\mathbb{Z}, +)$ and $(D_n, \circ)$ is that when adding integers, the order doesn't matter (that is, $a + b = b + a$ for any pair of integers), whereas, the order does matter when composing functions/symmetries. We give a special name to groups whose operation is commutative, that is, where the order does not matter.

Definition **3.10.** A group $(G, *)$ is **abelian** if and only if, for every pair of elements $g, h \in G$, $g * h = h * g$. So, a group is abelian if and only if its binary operation is commutative.

3.4. Subgroups, Generators, and Cyclic Groups

Cyclic groups give us examples of abelian groups.

Theorem 3.42. *If G is a cyclic group, then G is abelian.*

This theorem gives us an alternative method of seeing that the groups D_n are not cyclic.

Corollary 3.43. *The groups D_n for $n > 2$ are not cyclic.*

We have been exploring the relationship between groups that are cyclic and those that are abelian. We have seen that all cyclic groups are abelian, so we are left with the question of whether there are abelian groups that are not cyclic.

Exercise 3.44. 1. Give an example of an infinite group that is abelian but not cyclic.

2. Give an example of a finite group that is abelian but not cyclic. The smallest such group has four elements and is most easily described by writing its Cayley table.

In abelian groups every element commutes with every other element. In non-abelian groups, there can still be some elements that commute with all the elements of the group. We know that the identity element always commutes with every element, for example. We will name the set of elements that commute with every element of a group.

Definition 3.11. The **center** of a group G is the collection of elements in G that commute with all the elements of G. The center is denoted $Z(G)$ and can be described as

$$Z(G) = \{g \in G \mid g * h = h * g \text{ for all } h \in G\}.$$

The center of a group is not just a collection of elements of the group, it is a subgroup of the group.

Theorem 3.45. *Let G be a group. Then $Z(G)$ is a subgroup of G.*

The previous theorem is very exciting. We have found an interesting subgroup of a group without knowing anything else about it.

Exercise 3.46. Give examples of groups G in which

1. $Z(G) = \{e\}$;

2. $Z(G) = G$; and

3. $\{e\} \subsetneq Z(G) \subsetneq G$.

3.5 Sizes of Subgroups and Orders of Elements

The previous section set out to look for new examples of groups inside the examples already familiar to us. However, along the way the issue of size (including finite and infinite) appeared. In this section, we continue our investigation of arbitrary groups by making sense of the "size" of a group, a subgroup, and an element. In particular, we hope to understand any relationships between these quantities that are true in all groups.

We begin by breaking a group into pieces. If we want to understand the relationship between the size of a group and the size of a subgroup, one of those pieces will be a subgroup. But we saw above that some conditions don't give us a subgroup. For example $\{0, 3, 6, 9\}$ is a subgroup of $(C_{12}, \oplus_{12})$, but sets such as $\{1, 4, 7, 10\}$ and $\{2, 5, 8, 11\}$ are not subgroups even though they seem quite similar. These subsets seem similar because they are just shifted versions of the subgroup. The next definition captures this notion of a shift of a subgroup.

***Definition* 3.12.** Let H be a subgroup of a group G and $g \in G$. Then the **left coset of H by g** is the set of all elements of the form gh for all $h \in H$. This left coset is written $gH = \{gh \mid h \in H\}$. Right cosets are defined similarly.

The notation in the previous definition works well when the binary operation $*$ of the group G is written multiplicatively, like $g*h = gh$, for example in D_n or $(\mathbb{R} \setminus \{0\}, \cdot)$. But when the binary operation is written additively, with a plus sign, this notation can be confusing. So when writing the cosets of an additive group we use a $+$ notation for the cosets. For example, consider the group $(\mathbb{Z}, +)$. Then the cosets of the subgroup $3\mathbb{Z}$ are written $\{0 + 3\mathbb{Z}, 1 + 3\mathbb{Z}, 2 + 3\mathbb{Z}\}$.

***Exercise* 3.47.** 1. Consider

$$H = \{R_0, [\text{flip across a horizontal line}]\},$$

3.5. Sizes of Subgroups and Orders of Elements   63

a subgroup of D_4. Write out the left cosets of H. Also write out the right cosets of H.

2. Consider $K = \{[0]_{12}, [3]_{12}, [6]_{12}, [9]_{12}\}$, a subgroup of $\mathbb{Z}_{12}$. Write out the left cosets of K. Also write out the right cosets of K.

***Lemma* 3.48.** Let H be a subgroup of G and let g and g' be elements of G. Then the cosets gH and $g'H$ are either identical (the same subset of G) or disjoint.

Recall that $|G|$ denotes the order of the group G, that is, the number of elements in G.

***Lagrange's Theorem* 3.49.** Let G be a finite group with subgroup H. Then $|H|$ divides $|G|$.

***Scholium* 3.50.** Let G be a finite group with a subgroup H. Then the number of left cosets of H is equal to the number of right cosets of H.

Since the order of a subgroup divides the order of the group an integer number of times, it is natural to define a term that records that integer.

***Definition* 3.13.** Let H be a subgroup of a group G. Then the **index** of H in G is the number of distinct left (or right) cosets of H. We write this index as $[G : H]$.

***Scholium* 3.51.** Let G be a finite group with a subgroup H. Then $[G : H] = |G|/|H|$.

Lagrange's Theorem has many implications. One is that the order of each element must divide the order of the group.

***Corollary* 3.52.** Let G be a finite group with an element g. Then $o(g)$ divides $|G|$.

***Corollary* 3.53.** If p is a prime and G is a group with $|G| = p$, then G has no non-trivial subgroups.

3.6 Products of Groups

We've described many interesting groups, and we know that one way to find other groups is to find subgroups of the ones we have. Another method of producing new groups for the menagerie could come from combining (pairs of) examples that we already know.

Groups start out as sets, and some of you may be familiar with a way to take two sets and build a new, different one. For example, you are all familiar with the plane, $\mathbb{R}^2$. When we write $\mathbb{R}^2 = \mathbb{R} \times \mathbb{R}$, we see that the plane is built out of two copies of the familiar $\mathbb{R}$; and yet, $\mathbb{R}^2$ has exciting properties that are different from $\mathbb{R}$. We want to say that $\mathbb{R} \times \mathbb{R}$ is the product of two copies of $\mathbb{R}$.

Definition 3.14. If A and B are sets, then we define $A \times B = \{(a, b) \mid a \in A \text{ and } b \in B\}$, the set of ordered pairs of elements from A and B, which is called their **cartesian product**.

We can make the cartesian product of two groups into a group.

Theorem 3.54. Let $(G, *_G)$ and $(H, *_H)$ be groups and define

$$* : (G \times H) \times (G \times H) \to G \times H$$

by

$$(g_1, h_1) * (g_2, h_2) = (g_1 *_G g_2, h_1 *_H h_2).$$

Then $(G \times H, *)$ is a group, called the **(direct) product** of G and H.

The next exercise asks you to explore when the direct product of two cyclic groups is or is not a cyclic group.

Exercise 3.55. For natural numbers n and m, when is the group $\mathbb{Z}_n \times \mathbb{Z}_m$ cyclic?

The direct product of cyclic groups may not always be cyclic; however, the direct product of abelian groups is always abelian.

Theorem 3.56. Let G and H be groups. Then $G \times H$ is abelian if and only if both G and H are abelian.

The direct product of two groups has some natural subgroups.

Theorem 3.57. Let G_1 be a subgroup of a group G and H_1 be a subgroup of a group H. Then $G_1 \times H_1$ is a subgroup of $G \times H$.

3.7. Symmetric Groups

If we can realize a complicated group as the direct product of smaller groups, then we can feel that we know a lot about its structure. One of the most famous such structure theorems tells us that every finite abelian group is the direct product of cyclic groups.

Theorem* 3.58. Every finite abelian group is the direct product of cyclic groups of prime-power order.

In fact, this result can be extended to infinite abelian groups if they are finitely generated.

Theorem* 3.59. Every finitely generated abelian group is the direct product of some finite number of copies of $(\mathbb{Z}, +)$ and a finite number of cyclic groups of prime-power order.

These theorems are quite challenging at this stage of our development, but they represent a huge victory in our goal of describing the full potential captured in the definition of a group. Sadly, these results only hold for finitely generated abelian groups, so they are not complete.

3.7 Symmetric Groups

The previous section ended with the realization that we need more general examples of groups. As we've seen, one of the principal methods for developing mathematics is to modify and extend what we already know to create new examples and ideas. We have already employed this technique when we modified the example $\mathbb{Z}$ to create the examples C_n and $\mathbb{Z}_n$. Another example of this technique let us extend the example of rigid transformations of a triangle, namely the group D_3, to produce the related examples D_n, which are the rigid transformations of the regular n-gon. Here we will again start with the group D_3 and create new examples of groups, this time by concentrating on the representation of D_3 as a set of functions whose domain and range are the vertices of a triangle. The focus on functions will hopefully allow this new family of examples to be highly general.

Let's begin by establishing a new way of denoting the elements of D_3. Suppose we denote the three vertices by the numbers 1, 2, and 3 and we think of them as labeled around the triangle in a counterclockwise direction. Then we could denote the counterclockwise rotation of 120 degree by a two row matrix where each element in the top row of the notation is mapped to the element below it. This notation will be called the **two-line** notation.

$$g = \begin{pmatrix} 1 & 2 & 3 \\ 2 & 3 & 1 \end{pmatrix}$$

Or we can denote the same counterclockwise notation by (123), where this notation is interpreted to mean that each number goes to the number to its right, except the last number in the parenthesis, which goes to the first number in the parenthesis. This notation is called the **cycle** notation.

Exercise 3.60. 1. Write down a Cayley table for D_3 using the cycle notation for the elements of D_3.

2. Using cycle notation, illustrate that the group D_4 is not abelian.

Exercise 3.61.

1. The group D_3 is a collection of functions from the set $\{1, 2, 3\}$ to itself, with composition as the group operation. Could you construct a larger collection of functions from the set $\{1, 2, 3\}$ to itself that, again with composition as the group operation, would form a larger group? If so, describe the larger collection of functions. If not, why not.

2. The group D_4 is a collection of functions from the set $\{1, 2, 3, 4\}$ to itself, with composition as the group operation. Could you construct a larger collection of functions from the set $\{1, 2, 3, 4\}$ to itself that, again with composition as the group operation, would former a larger group? If so, describe the larger collection of functions. If not, why not.

3. For any natural number n describe the largest collection of functions from $\{1, 2, 3, \ldots, n\}$ to itself that would form a group under composition. That group is called the **symmetric group** on n elements and is denoted S_n.

4. Describe the two-line notation for elements of S_n.

3.7. Symmetric Groups

5. Describe the cycle notation for elements of S_n.

6. Describe the process by which you would compose two elements of S_n that are written in cycle notation. Create some examples to illustrate your method, perhaps in S_8.

Exercise 3.62. Write out the cycle notation for all elements of S_4.

Exercise 3.63. State and prove a conjecture of the form: The symmetric group on n elements, S_n, has _____ elements.

Exercise 3.64. Suppose $g \in S_n$ and that you know the cycle notation for g. How can you compute $o(g)$ without repeatedly composing g with itself?

Exercise 3.65. Find 30 distinct subgroups of S_4. (Hint: Each subgroup is generated by 1 or 2 elements.) Do you see subgroups of all orders that Lagrange's Theorem allows?

What properties of the functions in the groups D_n and S_n were necessary and what were superfluous for constructing a group of functions? Let's think about what properties of functions relate to the four properties of a group. First, the fact that the functions in D_n are from a set to itself makes composition sensible in all cases, so the binary operation is closed and well-defined. Associativity is automatic for composition of functions, so we won't need to worry about that. You may recall that a function needs to be injective to have an inverse under composition, and for that inverse to have the correct domain, the original function needs to be surjective. Putting these observations together, we should try to make a group whose elements are bijective functions from a set to itself.

Theorem 3.66. Let X be a set, let $Sym(X)$ be the set of bijections from X to X, and let $\circ$ represent composition. Then $(Sym(X), \circ)$ is a group, the **symmetric group** on X.

It turns out that *any* group can be thought of as a subgroup of a symmetric group. To prove this fact, our challenge is to associate an arbitrary element of an arbitrary group with a bijection on some set.

Exercise 3.67. Let G be a group. Find a set X and a natural injective function from G into $Sym(X)$. This injection allows us to recognize G as basically a subgroup of $Sym(X)$.

One of the strategies of mathematical exploration is to find the most general or most comprehensive examples of a mathematical object. The previous exercise suggests that understanding symmetric groups and their subgroups amounts to understanding *all* groups. So, this exercise is another major milestone on the way to classifying groups.

Unfortunately, another way to look at these insights is to say that the symmetric groups are as complicated as any groups that exist, so they will be difficult to fully fathom. In any case, thinking about elements of groups as permutations is often a valuable strategy. We'll talk more about this later, in the Section *Groups in Action*.

3.8 Maps between Groups

The previous section ended with the claim that every group is the same as a subgroup of a symmetric group. Moreover, the evidence for this claim came from a function that mapped one group into another. These claims make us want to know what "the same" means and how it can be captured with a function.

Our concept of "sameness" should depend on what we view as the fundamental, defining features of the groups. The definition tells the story: a group is a non-empty set together with a binary operation. So if we look at two groups, we want the concept of "sameness" to refer to the sets involved and their respective binary operations. Pinning this idea down is a basic strategy for exploring a mathematical idea. Once we have defined a mathematical object (in this case a group), we can ask what kind of functions between these objects (groups) respect the defining structure of the object (the sets and binary operations). Let's see what this abstract philosophy means in the case of groups.

For two groups G and H to be the same, their underlying sets should be in bijective correspondence. Thinking in terms of finite groups, there should be a relabeling of the elements that makes the Cayley tables look identical. Computing the binary operation before or after the relabeling should not matter. When two groups are the same in this sense that one group is just a relabeling of the elements of the other, then we call the groups *isomorphic*.

3.8. Maps between Groups

Exercise **3.68**. Using this informal definition of isomorphic, show that C_4 and $\mathbb{Z}_4$ are isomorphic.

The way we formalize the idea that two sets, X and Y, are relabelings of each other is by finding a bijection $f : X \to Y$. So for two groups G and H to be the same, there must be a bijection $\phi : G \to H$. But, in addition, the relabeling of the elements should respect the binary operations. Suppose that the elements a, b, c in G correspond respectively to the elements α, β, γ in the group H and in the group G and $a *_G b = c$, then we want $\alpha *_H \beta = \gamma$ in the group H. If $\phi : G \to H$ is the function that defines the relabeling, then we're saying that

$$\phi(a *_G b) = \phi(c) = \gamma = \alpha *_H \beta = \phi(a) *_H \phi(b).$$

A function from one group to another can respect the binary operations without necessarily being merely a relabeling, that is, without being a bijection. So now we seek to formalize what it means for a function, $\phi : (G, *_G) \to (H, *_H)$, to "respect the binary operations" regardless of whether ϕ is a bijection or not.

Definition **3.15**. Let $(G, *_G)$ and $(H, *_H)$ be groups and let $\phi : G \to H$ be a function on their underlying sets. Then we call ϕ a **homomorphism** of the groups if for every pair of elements $g_1, g_2 \in G$, $\phi(g_1 *_G g_2) = \phi(g_1) *_H \phi(g_2)$.

Notice that the central point in the definition of a homomorphism is that the binary operation in the domain is related to the binary operation in the range; first doing the binary operation on two elements in G and then performing the homomorphism gives the same element in H as first performing the homomorphism on the two elements individually and then combining the results with the binary operation in H. "Combine then map" should be the same as "map then combine".

Let's begin by examining a few homomorphisms between pairs of our favorite groups.

Exercise **3.69**. Confirm that each of the following functions is a homomorphism.

1. $\phi : \mathbb{Z}_{12} \to \mathbb{Z}_{24}$ defined by $\phi([a]_{12}) = [2a]_{24}$. (However, notice that $\phi([a]_{12}) = [a]_{24}$ is not a homomorphism.)

2. $\phi : \mathbb{Z} \to \mathbb{Z}_9$ defined by $\phi(a) = [3a]_9$

3. $\phi : \mathbb{Z}_6 \to \mathbb{Z}_3$ defined by $\phi([a]_6) = [a]_3$

4. $\phi : \mathbb{Z}_n \to \mathbb{Z}_n$ defined by $\phi([a]_n) = [-a]_n$

Definition 3.16. Let A be a subset of a set B. The **inclusion map** $i_{A \subset B} : A \to B$ is defined as follows: for each element $a \in A$, $i_{A \subset B}(a) = a$.

Theorem 3.70. Let H be a subgroup of a group G. Then the inclusion of H into G, $i_{H \subset G} : H \to G$, is a homomorphism.

Let's see some basic consequences of the definition of homomorphisms. The next theorem tells us that any homomorphism takes the identity of the domain group to the identity of the codomain group. If there are several groups floating around, we may write e_G for the identity of G.

Theorem 3.71. If $\phi : G \to H$ is a homomorphism, then $\phi(e_G) = e_H$.

Similarly, this next theorem tells us that homomorphisms send inverses to inverses.

Theorem 3.72. If $\phi : G \to H$ is a homomorphism and $g \in G$, then $\phi(g^{-1}) = [\phi(g)]^{-1}$.

These last two theorems suggest that homomorphisms send subgroups to subgroups. We need a little notation before we can state these observations carefully.

Definition 3.17. Let A and B be sets and $f : A \to B$ be a function. For any subsets $S \subset A$ and $T \subset B$ we define

1. $\text{Im}_f(S) = \{b \in B | \text{ there exists an } a \in S \text{ such that } f(a) = b\}$. We call $\text{Im}_f(S)$ the **image** of S (under f).

2. $\text{Preim}_f(T) = \{a \in A | f(a) \in T\}$. We call $\text{Preim}_f(T)$ the **preimage** of T (under f).

The set $\text{Im}_f(S)$ is the collection of elements in the codomain "hit" by elements in S; we often abuse notation and write $f(S) = \text{Im}_f(S)$. The biggest possible image, $\text{Im}_f(A)$, is then the set of elements in the

3.8. Maps between Groups

codomain that are hit by something, which is sometimes called the image of f, written as $\text{Im}(f)$. The set $\text{Preim}_f(T)$ is the collection of elements in the domain that "land" in T; we often abuse notation and write $f^{-1}(T)$ to mean the set $\text{Preim}_f(T)$. Note that f does not need to have an inverse for $f^{-1}(T)$ to be defined.

The next theorem tells us that the group structure is preserved by homomorphisms in the sense that the image of a group is a subgroup of the codomain.

Theorem 3.73. Let G and H be groups, let K be a subgroup of the group G, and let $\phi : G \to H$ be a homomorphism, then $\phi(K)$, which is also written as $\text{Im}_\phi(K)$, is a subgroup of H.

Corollary 3.74. If ϕ is a homomorphism from the group G to the group H, then $\text{Im}(\phi)$, which is also written as $\text{Im}_\phi(G)$ and as $\phi(G)$, is a subgroup of H.

Exercise 3.75. Let G be a group with an element g. Then define the function $\phi : \mathbb{Z} \to G$ by setting $\phi(n) = g^n$ for each $n \in \mathbb{Z}$. Show that ϕ is a homomorphism. We have seen $\text{Im}(\phi)$ before; what is another name for this image subgroup?

Exercise 3.76. Let $G = \langle g \rangle$ be a cyclic group and $\phi : G \to H$ be a homomorphism. Show that knowing $\phi(g)$ allows you to compute $\phi(g')$ for all $g' \in G$.

The subgroup-preserving property of images of homomorphisms also works with the preimages of subgroups of the codomain under a homomorphism.

Theorem 3.77. Let G and H be groups, let L be a subgroup of the group H, and let $\phi : G \to H$ be a homomorphism, then $\phi^{-1}(L)$, which is also written as $\text{Preim}_\phi(L)$ and as $\{g \in G | \phi(g) \in L\}$, is a subgroup of G.

One such preimage is so important that it has a name of its own.

Definition 3.18. Let $\phi : G \to H$ be a homomorphism from a group G to a group H. Then the set $\{g \in G | \phi(g) = e_H\}$ is called the **kernel of** ϕ and is denoted $\text{Ker}(\phi)$.

Corollary 3.78. Let G and H be groups. For any homomorphism $\phi : G \to H$, $\text{Ker}(\phi)$ is a subgroup of G.

We are all familiar with several special types of functions: injective (1-1) functions, surjective (onto) functions, and bijective (1-1 and onto) functions. In Exercise **3.69** above, we saw examples of homomorphisms which, as functions, fell into each of these categories. We give special names to each of these types of homomorphisms.

Definitions **3.19.** 1. An injective homomorphism is called a **monomorphism**.

2. A surjective homomorphism is called an **epimorphism**.

3. A bijective homomorphism is called an **isomorphism**.

4. A group G is **isomorphic** to a group H if there exists an isomorphism, $\phi : G \to H$. If G is isomorphic to H, we write $G \cong H$.

To get accustomed to these terms, let's begin by classifying each homomorphism from Exercise **3.69** above as a monomorphism, epimorphism, or isomorphism.

Exercise **3.79.** Classify each of the following homomorphisms as a monomorphism, an epimorphism, an isomorphism, or none of these special types of homomorphisms.

1. $\phi : \mathbb{Z}_{12} \to \mathbb{Z}_{24}$ defined by $\phi([a]_{12}) = [2a]_{24}$.

2. $\phi : \mathbb{Z} \to \mathbb{Z}_9$ defined by $\phi(a) = [3a]_9$

3. $\phi : \mathbb{Z}_6 \to \mathbb{Z}_3$ defined by $\phi([a]_6) = [a]_3$

4. $\phi : \mathbb{Z}_n \to \mathbb{Z}_n$ defined by $\phi([a]_n) = [-a]_n$

The integers can map onto any one of the modular arithmetic groups, $\mathbb{Z}_n$, by a homomorphism.

Exercise **3.80.** For any natural number n, there is an epimorphism $\phi_n : \mathbb{Z} \to \mathbb{Z}_n$.

Projections from a direct product of groups to one of the factors are examples of epimorphisms. Let's define our terms.

Definition **3.20.** Let X and Y be sets with cartesian product $X \times Y$. The functions $\pi_X((x, y)) = x$ and $\pi_Y((x, y)) = y$ are called **projections** to the first and second coordinates respectively.

3.8. Maps between Groups

Theorem 3.81. Let G and H be groups. Then the projection maps $\pi_G : G \times H \to G$ and $\pi_H : G \times H \to H$ are epimorphisms.

The next theorem relates homomorphisms into a pair of groups with a homomorphism into their direct product.

Theorem 3.82. Let G, H, and K be groups with homomorphisms $f_1 : K \to G$ and $f_2 : K \to H$. Then there is homomorphism $f : K \to G \times H$ such that $\pi_G \circ f = f_1$ and $\pi_H \circ f = f_2$. Furthermore, f is the only function satisfying these properties. Moreover, if either f_1 or f_2 is a monomorphism, then f is also a monomorphism.

The concept of isomorphism is extremely important because two groups being isomorphic captures the idea that the two groups are the "same" by formalizing the notion that the two groups are just relabelings of each other. We need to know when two groups are the same if we want to claim that we have classified all groups. We have already been introduced to two groups that should be the same: the cyclic arithmetic and modular arithmetic groups of the same order. Let's confirm this feeling by showing that they are isomorphic.

Theorem 3.83. For every natural number n, the two groups $(C_n, \oplus_n)$ and $(\mathbb{Z}_n, \oplus)$ are isomorphic.

After this theorem, we can stop being so careful about our notation when dealing with these cyclic groups. Since they are isomorphic, any purely group-theoretic question asked of them will give identical answers. We will use whichever version of the group lends itself to the question at hand, which is usually $\mathbb{Z}_n$.

We defined $G \cong H$ if there is a function $\phi : G \to H$ that is both bijective and a homomorphism. But perhaps we should have demanded that "the same" also required that there was a bijective homomorphism $\overline{\phi} : H \to G$. Fortunately, the bijection ϕ does have an inverse, $\phi^{-1} : H \to G$. The next theorem tells us that ϕ^{-1} is also a homomorphism. So, we could have defined an isomorphism as a bijection that respects the group operations in *both* directions.

Theorem 3.84. Let $\phi : G \to H$ be an isomorphism. Then $\phi^{-1} : H \to G$ is an isomorphism.

This previous theorem says that an isomorphism really is a relabeling of the group elements so that the Cayley tables look identical, as we had desired. Since an isomorphism between two groups means they are the same, we should check that the term behaves appropriately.

Theorem 3.85. Let G, H, and K be groups. Then

1. $G \cong G$;
2. if $G \cong H$ and $H \cong K$, then $G \cong K$; and
3. if $G \cong H$, then $H \cong G$.

In other words, "isomorphic" is an *equivalence relation*.

In the previous theorem, we proved that every group is isomorphic to itself. You probably saw that your group was isomorphic to itself using an uninteresting function. But there are often other ways that a group is isomorphic to itself. Such an isomorphism would need to mix up the elements of the group. The following theorem recalls a technique for mixing up the elements of a group that you may have noticed when working with cosets.

Theorem 3.86. Let G be a group with an element g. Define $\phi_g : G \to G$ by $\phi_g(h) = ghg^{-1}$. Then $\phi_g : G \to G$ is an isomorphism, called **conjugation by** g.

One way to tell whether a homomorphism is an isomorphism is to look at its kernel and its image. The next theorem tells us that it is enough to check that e_H has only one preimage under $\phi : G \to H$ to know that the whole function is injective.

Theorem 3.87. Let $\phi : G \to H$ be a homomorphism. Then ϕ is a monomorphism if and only if $\text{Ker}(\phi) = \{e_G\}$. In particular, ϕ is an isomorphism if and only if $\text{Im}(\phi) = H$ and $\text{Ker}(\phi) = \{e_G\}$.

The modular arithmetic groups give us many good examples of homomorphisms and isomorphisms.

Theorem 3.88. The map $\phi : \mathbb{Z}_n \to \mathbb{Z}_n$ defined by $\phi([a]_n) = [n-a]_n$ is an isomorphism.

Theorem 3.89. Let k and n be natural numbers. The map $\phi : \mathbb{Z}_n \to \mathbb{Z}_n$ defined by $\phi([a]_n) = [ka]_n$ is a homomorphism.

3.8. Maps between Groups

***Exercise* 3.90.** Make and prove a conjecture that gives necessary and sufficient conditions on the natural numbers k and n to conclude that $\phi : \mathbb{Z}_n \to \mathbb{Z}_n$ defined by $\phi([a]_n) = [ka]_n$ is an isomorphism. Use this insight to show that there are several different isomorphisms $\phi : \mathbb{Z}_{12} \to \mathbb{Z}_{12}$.

***Exercise* 3.91.**   1. Use conjugation to find two different isomorphisms from D_4 to D_4.

2. Why does conjugation not give any interesting isomorphisms from $\mathbb{Z}_n$ to itself?

The isomorphisms above are all straightforward, either from a group to itself or between groups obviously based on the same set (like C_n and $\mathbb{Z}_n$). But isomorphisms need not be so obvious.

***Exercise* 3.92.** Let $H = \{R_0, R_{120}, R_{240}\}$, which is a subgroup of D_3. Then H is isomorphic to the cyclic group C_3.

***Theorem* 3.93.** For each n, the symmetries of a regular n-gon, D_n, has a subgroup isomorphic to C_n.

One strategy of mathematical exploration is to find the most general or the most comprehensive examples of a mathematical object and study it. The previous theorem says that the symmetry groups of the regular polygons contain the finite cyclic groups as subgroups. Similarly, the symmetric groups contain the symmetry groups of the regular polygons.

***Theorem* 3.94.** For any n, the symmetric group S_n has subgroups isomorphic to D_n and C_n.

The previous theorem is a little misleading, since it makes it seem like only the groups corresponding to the same n have any relationships.

***Exercise* 3.95.** Find 40 different subgroups of S_6 isomorphic to C_3.

***Exercise* 3.96.** For any $m > n$, find a monomorphism $\phi : S_n \to S_m$.

And the crowning theorem tells us that *every* group is a subgroup of a symmetric group.

Theorem 3.97. Let G be a group. Then for some set X, there is a subgroup H of $Sym(X)$ such that G is isomorphic to H. If G is finite, then X can be chosen to be finite.

This last theorem classifies all groups, which was our goal! Sadly, the result is not very descriptive. So we continue our journey.

3.9 Normal Subgroups and Quotient Groups

In the previous section, we explored the maps between groups called homomorphisms, and we saw that each homomorphism gave us a special subgroup called the kernel. The elements in the kernel of a homomorphism all get "crushed" together by the homomorphism, and the result is the image, which is still a group. This suggests that we could describe the structure of an arbitrary group using kernels and images. This section asks what is special about kernels.

In general, a left coset gH may or may not be equal to the right coset Hg.

Exercise 3.98. Find a subgroup H of D_3 and an element g of D_3 such that gH is not equal to Hg.

Although in general a left coset gH may or may not be equal to the right coset Hg, when K is the kernel of a homomorphism, gK always equals Kg.

Theorem 3.99. Let G and H be groups and let $\phi : G \to H$ be a homomorphism. Then for every element g of G, $g\text{Ker}(\phi) = \text{Ker}(\phi)g$.

It is useful to give a name to those subgroups, like kernels of homomorphisms, with the property that each right coset is equal to the corresponding left coset.

Definition 3.21. A subgroup K of a group G is **normal**, denoted $K \triangleleft G$, if and only if for every element g in G, $gK = Kg$ or, equivalently, $K = gKg^{-1}$. (The final set is shorthand: $gKg^{-1} = \{gkg^{-1} | k \in K\}$.)

We can reformulate the previous theorem using this new vocabulary: if $\phi : G \to H$ is a homomorphism, then $\text{Ker}(\phi)$ is a normal subgroup of G. Moreover, preimages of normal subgroups are normal.

3.9. Normal Subgroups and Quotient Groups

Theorem 3.100. Let G and H be groups, $\phi : G \to H$ be a homomorphism, and K be a normal subgroup of H. Then $\phi^{-1}(K)$, which is also written as $\text{Preim}_\phi(K)$, is a normal subgroup of G.

In abelian groups, all subgroups are normal.

Theorem 3.101. Let K be a subgroup of an abelian group G. Then K is a normal subgroup.

An equivalent characterization of normal subgroups is often useful. Recall the definition of conjugation by g above. The reformulated characterization in the next theorem is often called "K is closed under conjugation by every element $g \in G$".

Theorem 3.102. A subgroup K of a group G is normal if and only if for every $g \in G$ and $k \in K$, $gkg^{-1} \in K$.

Recall that the center of a group G, $Z(G)$, is the set of all the elements of G that commute with every element of G. The center of a group is always a normal subgroup.

Theorem 3.103. Let G be a group. Then $Z(G) \triangleleft G$.

Now we understand that the cosets of a kernel (or any normal subgroup) are particularly well behaved. This inspires another technique for building new groups out of the old: use the cosets of a normal subgroup as the elements of a new group. Then the hard work becomes finding an appropriate binary operation. This strategy, which the following theorem describes, produces a group called a quotient group.

Theorem 3.104. Let K be a normal subgroup of a group $(G, *)$, and let G/K be the left cosets of K in G. Define the binary operation $\hat{*}$ on G/K by $gK \mathbin{\hat{*}} g'K = (g * g')K$. Then $(G/K, \hat{*})$ is a group, and $|G/K| = [G : K]$.

When K is a normal subgroup of G, the group G/K described above is called a **quotient group** and read "G mod K".

Exercise 3.105. Explain the necessity of the normality hypothesis in the definition of quotient groups. Give an example of a group G with a subgroup H such that the cosets of H do not form a group using the operation defined in Theorem **3.104**.

Let's look at an example of a quotient group. Consider the quotient group $\mathbb{Z}/3\mathbb{Z}$, whose elements are the cosets of the normal subgroup $3\mathbb{Z}$ of $\mathbb{Z}$. Those cosets are written $\{0+3\mathbb{Z}, 1+3\mathbb{Z}, 2+3\mathbb{Z}\}$. In $\mathbb{Z}/3\mathbb{Z}$, $(1+3\mathbb{Z})\hat{+}(1+3\mathbb{Z}) = (1+1)+3\mathbb{Z} = 2+3\mathbb{Z}$.

Theorem 3.106. For any natural number n, $n\mathbb{Z}$ is a normal subgroup of $\mathbb{Z}$ and $\mathbb{Z}/n\mathbb{Z} \cong C_n \cong \mathbb{Z}_n$.

To get accustomed to quotient groups, let's look at the quotient groups that arise from S_4.

Exercise 3.107. Find all normal subgroups K of S_4. For each such K show that S_4/K is isomorphic to $\{e\}$, C_n, D_n, or S_n for n equal to 2, 3, or 4.

Sometimes the structure of the quotient group can give us information about the whole group.

Theorem 3.108. Suppose that $G/Z(G)$ is a cyclic group. Then G is abelian. (So $Z(G) = G$.)

Conversely, sometimes knowledge about the normal subgroup gives us information about the character of the quotient group.

Theorem 3.109. Let K be a normal subgroup of a group G. Then G/K is abelian if and only if K contains $\{ghg^{-1}h^{-1} | g, h \in G\}$.

Because of its role in making the quotient operation commutative, the subgroup of G generated by elements of the form $ghg^{-1}h^{-1}$, that is

$$[G, G] = \langle\{ghg^{-1}h^{-1} | g, h \in G\}\rangle,$$

is called the **commutator subgroup of** G.

Theorem 3.110. Let G be a group, then $[G, G]$ is a normal subgroup of G. Thus $G/[G, G]$ is an abelian group and it is called the **abelianization** of G.

The previous theorem describes a method for turning an arbitrary group into an abelian group. The theorem implies that $G/[G, G]$ is the biggest possible image of G in an abelian group.

We are finally ready to stop worrying about the difference between a normal subgroup and a kernel: every normal subgroup is actually the

3.9. Normal Subgroups and Quotient Groups

kernel of a homomorphism. This result will help us with our goal of building groups out of kernels and images.

***Theorem* 3.111.** Let G be a group and K be a normal subgroup of G. Then there is an epimorphism $q : G \to G/K$ with $\text{Ker}(q) = K$. So a subgroup is normal if and only if it is the kernel of a homomorphism.

One of the most useful and important theorems in group theory relates the image of a homomorphism with a quotient group. The first step is to associate each element of the image of a homomorphism with a coset of the kernel of the homomorphism in a natural way. The following theorem specifies this association.

***Theorem* 3.112.** Let $\phi : G \to H$ be an epimorphism from a group G to a group H and let $h \in H$, then $\phi^{-1}(h) = g\text{Ker}(\phi)$ for some $g \in G$.

Using this association, you can state and prove one of the most fundamental theorems in group theory.

***First Isomorphism Theorem* 3.113.** Let G and H be groups. For any homomorphism, $\phi : G \to H$, the image of ϕ is isomorphic to the quotient group $G/\text{Ker}(\phi)$, or using the notation for isomorphism: $\text{Im}(\phi) \cong G/\text{Ker}(\phi)$.

***Corollary* 3.114.** Let G and H be groups. For any epimorphism, $\phi : G \to H$, $H \cong G/\text{Ker}(\phi)$.

The First Isomorphism Theorem allows us to determine the structure of many groups and their subgroups. See whether you can use the First Isomorphism Theorem to prove the following theorems.

***Theorem* 3.115.** For each natural number, n, there is a unique cyclic group with order n.

***Theorem* 3.116.** Let m and n be relatively prime integers. Then $\mathbb{Z}_m \times \mathbb{Z}_n \cong \mathbb{Z}_{mn}$.

As we mentioned at the end of the section on products, one of the central theorems in the study of abelian groups relates them to products of cyclic groups. The proof of the following theorem is rather involved, but can be done by finding an epimorphism from $\mathbb{Z}^n$ (= $\mathbb{Z} \times \cdots \times \mathbb{Z}$ n times) to the abelian group and cleverly describing the kernel.

Lemma 3.117. Let G be a finitely generated abelian group, generated by n elements $\{g_1, g_2, \ldots, g_n\}$. Then $\phi : \mathbb{Z}^n \to G$ defined by $(a_1, a_2, \ldots, a_n) \mapsto g_1^{a_1} g_2^{a_2} \cdots g_n^{a_n}$ is an epimorphism.

If we understood the kernel of the epimorphism from the previous lemma, we could use the First Isomorphism Theorem to prove the next theorem, called the Fundamental Theorem of Finitely Generated Abelian Groups. However, even with this good strategy firmly in mind, the following is a difficult theorem.

Fundamental Theorem of Finitely Generated Abelian Groups* 3.118. Any finitely generated abelian group is isomorphic to a direct product of cyclic groups.

This theorem is considered fundamental because it describes the structure of every finitely generated abelian group. Recognizing that any finitely generated abelian group is really just a product of cyclic groups shows us the basic structure and simplicity of those abelian groups. We have finally given a descriptive classification of a huge collection of groups.

3.10 More Examples*

Before we conclude our tour of group theory, let's describe a few additional groups. Like the symmetric groups, each element of the following group is a function.

Theorem 3.119. The set $M = \{f : \mathbb{R} \to \mathbb{R} |$ such that $f(x) = \frac{ax+b}{cx+d}$, where $a, b, c, d \in \mathbb{R}$ and $ad - bc = 1\}$ with the binary operation of composition forms a group.

Here is a group whose elements are matrices.

Theorem 3.120. The set of 2×2 matrices with real number entries and determinant equal to 1, written $SL_2(\mathbb{R})$, along with the operation of matrix multiplication is a group.

3.11. Groups in Action*

***Theorem* 3.121.** The groups M and $SL_2(\mathbb{R})/\langle \begin{pmatrix} -1 & 0 \\ 0 & -1 \end{pmatrix} \rangle$ are isomorphic.

***Exercise* 3.122.** Realize D_4 as a subgroup of $SL_2(\mathbb{R})$ by writing each transformation as a matrix. A good strategy is to find an injective homomorphism $\phi : D_4 \to SL_2(\mathbb{R})$. The First Isomorphism Theorem will then establish that the image of ϕ is the desired subgroup.

The self-isomorphisms of a group form a group of their own.

***Definition* 3.22.** Let G be a group. Then $Aut(G)$ is the set of isomorphisms from G to itself, which comes with composition of functions as a binary operation. The elements of $Aut(G)$ are called **automorphisms** of G.

***Theorem* 3.123.** Let G be a group. Then $(Aut(G), \circ)$ is a subgroup of $Sym(G)$. In particular, $Aut(G)$ is a group.

***Theorem* 3.124.** If p is a prime number, then $Aut(\mathbb{Z}_p) \cong \mathbb{Z}_{p-1}$.

We've actually already seen some automorphisms. Let g be an element of the group G. Then we defined $\phi_g : G \to G$, conjugation by g, by $\phi_g(h) = ghg^{-1}$ for every $h \in G$.

***Theorem* 3.125.** Let G be a group. For each element $g \in G$, ϕ_g is an automorphism of G. Furthermore, $\Phi : G \to Aut(G)$ defined by $\Phi(g) = \phi_g$ is a homomorphism.

The preceding theorem provides us with a bunch of automorphisms for non-abelian groups, but for abelian groups conjugation is just the trivial automorphism. The next theorem tells us that every group with at least three elements has at least one non-trivial automorphism.

Theorem * **3.126.** Let G be a group with more than 2 elements. Then $|Aut(G)| > 1$.

3.11 Groups in Action*

Although we introduced groups as abstract structures, they actually appear in many different applications. In fact, historically, groups grew out of observations about collections of permutations. These groups that permute things can be thought of as *acting on a set*. We've already seen a few groups that act on sets in that sense, for example, D_n transforms the regular n-gon, and $Sym(X)$ permutes the elements of X. Since $Sym(X)$ includes *any* permutation of the set X, it is our most general example of a group acting on a set. When thinking about a group acting on a set, we want the elements of the group to be associated with permutations, but in a way that respects the group structure. We'll first give a formal definition of this idea and then explain what it means.

Definition 3.23. Let G be a group. Then an **action** of G on a set X is a homomorphism $\phi : G \to Sym(X)$. We say that G *acts on* X by ϕ.

At first, the term "action" might seem a little confusing. However, a map ϕ from G to $Sym(X)$ allows us to associate each element g of G with a permutation, namely the permutation $\phi(g) \in Sym(X)$. With that relationship, G is associated with a collection of permutations of the set X. Using this association, we will be able to employ insights concerning permutations to answer group-theoretic questions.

We have already seen several examples of groups acting on sets. In proving Theorem **3.97**, you probably used an insight about groups that can now be stated in terms of a group acting on itself by left multiplication.

Theorem 3.127. Let G be a group. For each $g \in G$, define $\lambda_g : G \to G$ by $\lambda_g(h) = gh$. Then $\Lambda : G \to Sym(G)$ defined by $g \mapsto \lambda_g$ is an action of G on G.

Similarly, G acts on the cosets of a subgroup H.

Theorem 3.128. Let H be a subgroup of a group G and let $L = \{gH | g \in G\}$ be the set of left cosets of H. Then the function $\phi : G \to Sym(L)$ defined by $\phi(g)(g'H) = (gg')H$ is an action of G on L.

We have already seen a second example of a group G acting on itself in Theorem **3.125** when we defined the homomorphism $\Phi : G \to Aut(G) \subset Sym(G)$ that associated each element g of G with the automorphism: conjugate by g. In other words, G acts on itself by conjugation.

3.11. Groups in Action*

When we begin to explore the idea of group actions, two ideas arise about how the elements of G are moving the elements of X around. The first natural question is, "For each element of X, what elements of G leave it fixed?". The second natural question is, "Where does an element $x \in X$ go under the permutations of G?". These questions lead to two definitions.

Definitions 3.24. 1. Let G be a group with an action $\phi : G \to Sym(X)$ and let $x \in X$. The set of group elements that fix x, called the **stabilizer** of x, is $Stab(x) = \{g \in G \,|\, \phi(g)(x) = x\}$.

2. The **orbit** of x is $Orb(x) = \{y \in X \,|\, y = \phi(g)(x) \text{ for some } g \in G\}$, which is just the collection of elements that x gets mapped to by the action.

Exercise 3.129. Pick a non-trivial subgroup H of D_4, and consider D_4 acting on the left cosets of H by left multiplication. For each coset gH, find its stabilizer and orbit under this action.

The reason for requiring that an action of G on X be a homomorphism from G to $Sym(X)$, instead of just any old function, is that we would like the action of an element g followed by the action of an element h to be the same as the action of the element hg. If we write out this condition, it is precisely what is required for the action to be a homomorphism. This condition guarantees that stabilizers are subgroups. In other words, an action respects the group's structure.

Theorem 3.130. Let G be a group and $\phi : G \to Sym(X)$ be an action of G on X. If $x \in X$, then $Stab(x)$ is a subgroup of G.

One of the neatest features about group actions is that there is a basic relationship between the number of places an element of X goes to under the action of G and the number of elements of G that leave it fixed.

Theorem 3.131. Let G be a finite group acting on a set X. Then for any $x \in X$,
$$|Stab(x)| \cdot |Orb(x)| = |G|.$$

The next lemma says that the orbits partition X. This should remind you of the proof that the cosets of a subgroup partition the group. Even the proofs are similar.

Lemma 3.132. Let G be a group acting on a set X. Then for two elements $x, y \in X$, either $Orb(x) = Orb(y)$ or they are disjoint.

Theorem 3.133. Let G be a finite group acting on a finite set X. Then

$$|X| = \sum \frac{|G|}{|Stab(x_i)|},$$

where the sum is taken over one element x_i from each distinct orbit.

These theorems that relate the sizes of the group, the set, the stabilizers, and the orbits give clever methods for gaining insights into the structure of finite groups. For example, Cauchy proved that if a prime p divides the order of a group, then the group has an element of order p. Let's prove it using a group action.

Lemma 3.134. Let G be a finite group. Let $E(n)$ be the set of all n-tuples of elements of G such that the product of those n elements in order equals the identity element. That is,

$$E(n) = \{(g_1, g_2, g_3, \ldots, g_n) | g_i \in G \text{ and } g_1 g_2 g_3 \ldots g_n = e_G\}.$$

Then $|E(n)| = |G|^{n-1}$.

Lemma 3.135. Let G be a finite group, n be a natural number, and $E(n) = \{(g_1, g_2, g_3, \ldots, g_n) | g_i \in G \text{ and } g_1 g_2 g_3 \ldots g_n = e_G\}$. Let $\phi : \mathbb{Z}_n \to Sym(E(n))$ be defined by cyclic permutation, that is,

$$\phi([i]_n)((g_1, g_2, g_3, \ldots, g_n)) = (g_{1+i}, g_{2+i}, \ldots, g_{n+i})$$

where the subscripts are interpreted mod n. Then

$$Stab((e_G, e_G, e_G, \ldots, e_G)) = \mathbb{Z}_n.$$

Cauchy's Theorem 3.136. Let G be a finite group and p be a prime that divides the order of G, then G has an element of order p.

Recall that Lagrange's Theorem stated that the order of any subgroup of a finite group divides the order of the group. Cauchy's Theorem is a partial converse to Lagrange's Theorem. A more substantial (but still partial) converse due to Sylow can also be proved using the ideas of groups acting on sets. Cauchy's Theorem is a special case of this more general theorem.

3.12. The Man Behind the Curtain

Sylow's Theorem* 3.137. Let G be a finite group and let p be a prime. If p^i divides the order of G, then G has a subgroup of order p^i.

One of the most fruitful sets on which a group can act is itself. We have already seen the action of conjugation, and we will now look at some further consequences of that action. This action is so important that its orbits and stabilizers have special names.

Definitions 3.25. 1. Let G be a group. For any element $g \in G$, define $C_G(g) = \{h \in G | hgh^{-1} = g\}$, called the **centralizer** of g.

2. Let g be an element of G; then the **conjugacy class** of g is the set $\{hgh^{-1} | h \in G\}$.

Exercise 3.138. Describe the conjugacy classes in the symmetric groups, S_n.

Corollary 3.139. Let G be a finite group. Then

$$|G| = \sum [G : C_G(g_i)],$$

where the sum is taken over one element, g_i, from each conjugacy class in G.

This last corollary can be used to show that certain groups have non-trivial centers, without producing specific elements that commute! The proof is direct but not constructive, which makes it a rare and unsettling beast.

Theorem 3.140. Let G be a group such that $|G| = p^n$ for some prime p. Then $|Z(G)| \geq p$.

3.12 The Man Behind the Curtain

Many people mistakenly believe that mathematics is arbitrary and magical, or at least that there is some secret knowledge that math teachers have but won't share with their students. Mathematics is no more magical than the Great and Powerful Wizard of Oz, who was just a man behind a curtain. The development of mathematics is directed by a few simple principles and a strong sense of aesthetics. To develop the ideas of group theory we followed a path of guided discovery. Let's look back on the journey and let the guiding strategies emerge from behind the curtain.

Group theory has many applications, ranging from internet credit card security and cracking the Enigma Code to solving the Rubik's Cube and generalizing the quadratic formula. Our introduction to group theory has only exposed us to a tiny portion of the theorems living in the wild.

We started with familiar activities: adding, multiplying, telling time, and putting blocks in a matching hole. The definition of a group emerged by distilling the essential features and commonalities from these specific examples in the process that we call abstraction. These examples generalized to produce cyclic groups, the dihedral groups, and symmetric groups in the process we call exploration. We defined generators, subgroups, products, and homomorphisms to help us discover and describe the wide variety of groups that we caught in our definition-net. We saw patterns that led us to conjectures about the size of subgroups; we justified these insights, producing theorems. We applied those insights to extend our exploration of the concept of groups.

This investigation represents a complete cycle of inquiry, and it especially emphasizes and deepens the skill of definition exploration. We explored a mathematical idea in its own right, hoping to describe the rich mosaic of possibilities that an abstract definition could capture. In particular, trying to *classify* all groups was one of the motivations that guided our exploration. Let us reflect on these new tools in our inquiry tool belt.

- **Generation** – After choosing the definition of a group, we explored its potential by building new examples. We modified the motivating examples and generalized them to families of examples including the dihedral and symmetric groups (**3.12**, **3.13**, Definition **3.3**, **3.19**, **3.54**, **3.66**, **3.119**, **3.120**, **3.123**), described properties that distinguish examples from one another (properties such as being abelian or cyclic) (**3.4**, **3.34**. Definition **3.10**, **3.43**, **3.44**, **3.46**), articulated the ways in which the examples can be generated by their elements and explored the examples generated by

3.12. The Man Behind the Curtain

one or a small number of elements (**3.27**, **3.29**), defined a notion of size including finite and infinite (Definition **3.8**, Definition **3.9**), created new examples by combining existing examples including creating product groups (**3.54**), created new examples by grouping or identifying elements including creating quotient groups (**3.104**), defined sub-objects and listed all examples found inside the known examples (Definition **3.4**, **3.22**), asked if there could be biggest and smallest examples (**3.24**), and found ways to transform groups into related groups through processes including abelianization (**3.110**).

- **Structural Investigation** – We also turned our magnifying lens toward the general definition to uncover its hidden consequences. We asked if specified structures like the identity and inverses might actually be unique (**3.6**, **3.9**), asked if the specified conditions might be stronger than needed as with two-sided inverses (**3.10**), listed special subgroups like the trivial subgroups, the center, and subgroups generated by individual elements (**3.24**, **3.25**, **3.27**, Definition **3.11**), and articulated the ways in which examples are indecomposable (Definition **3.6**).

- **Identification** – We explored the relationship between pairs of examples by defining the notion of "sameness" for two objects using functions (isomorphisms) (Definition **3.18**), generalizing those functions to the maps that preserve structure (homomorphisms) (Definition **3.15**), exploring the sub-objects that arise as images and preimages of other sub-objects (**3.73**, **3.78**), and exploring the ways in which a single example can be the same as itself in non-trivial ways (automorphisms) (**3.86**, Definition **3.22**). After a mathematical structure has been defined, it is natural to identify those functions (such as homomorphisms, isomorphisms, and automorphisms) that respect the structure being explored. The pattern of defining "sameness" through functions that preserve structure occurs in many branches of mathematics.

- **Classification** – Finally we asked if the possible examples can be classified either by exhibiting them all as sub-objects of one mega-object (**3.67**) or by describing how they can all be built from the simplest examples in a predictable fashion (**3.58**, **3.59**, **3.118**). It is a rare and exciting mathematical moment when classification is possible.

The inquiry tools described here are not sequential; we moved among them repeatedly. Example generation, questions of uniqueness, a definition of isomorphism, and some classification are common strategies for exploring mathematical objects. Consider looking back on your previous mathematical experiences for more examples of these strategies in action. Then use these techniques to help you explore the wilds during your future mathematical adventures.

4

Calculus

4.1 Perfect Picture

The Summer Olympic Games are contested every four years. Many modern sports are exciting, but the traditional ones harken back to the ancient roots of the games. When a Greek contender is among the world's best at archery, you can just imagine the excitement in the air. Camera crews were poised to record the thrilling finish as Zeno stepped up to the line and drew his bow. His release was smooth and apparently effortless. The arrow flew toward the target and the world's eyes were glued to their television sets as the arrow neared the winning bull's eye. Unfortunately, at the very instant when the arrow would have hit the target, the electricity went out. Even worse, at the moment of impact, the target, which had been constructed by the lowest bidder and was made of compressed sawdust, simply disintegrated completely. The world gasped in anguish as the judges were deposited in an immediate quandary—what to do?

Following the practice of the NFL, they decided to use a video review. The two lead referees were named Isaac and Gottfried. These judges were experts, but the problem was that the ultimate evidence—the arrow touching the target—did not exist. The videotape clearly showed the location of the arrow at many moments as it approached the target; however, the tape did not include that last moment of actual touching.

Many of the lesser judges, particularly those being paid off by the countries of the other contenders, argued that without evidence of the arrow actually touching the bull's eye, the Gold Medal could not be awarded to Zeno. However, those lesser judges were no match for the decisive arguments put forth by Isaac and Gottfried. Here is a transcript of their convincing arguments.

Referee Isaac: At 12:59:59 PM Zeno's arrow was 100 ft. from the target, as seen in this photo.

Referee Gottfreid: At 12:59:59.5 PM Zeno's arrow was 25 ft. from the wall, as seen in this second photo.

Referee Isaac: At 12:59:59.75 PM Zeno's arrow was $\frac{25}{4}$ ft. from the wall, as seen in this third photo.

Referee Gottfreid: In fact, at $(\frac{1}{2})^n$ seconds before 1:00 PM, Zeno's arrow was $\frac{100}{4^n}$ ft. from the wall. This evidence involves an infinite number of photographs, but we had a special camera that could take this infinite sequence of pictures in the moments leading up to 1:00 PM. (It's no surprise that this camera sucked all of the electricity out of the power grid and stopped working right when it did, huh?)

Referee Isaac: Although we do not have a photograph at 1:00 PM showing the arrow actually touching the target, these photos show that the arrow must have made contact with the target at 1:00 PM because the positions of the arrow at times arbitrarily close to 1:00 PM converge to touching the bull's eye at 1:00 PM.

Zeno: Great! But converge? What does *converge* mean?

Referee Gottfried: Just think it over for 150 years and you'll understand. There will be plenty of time while posterity celebrates your Gold Medal victory through the ages.

The moral of this story is that the motion of an object in space is predictable; objects do not jump or teleport when moving. If you know the position of Zeno's arrow except at one instant in time, you know where it is at that time as well.

Suppose you know the position of a certain particle at all times right before and right after time t_0, but not right at t_0. You can think

4.2. Convergence

of this scenario as a movie that is missing a single frame in the middle. There is only one way to insert the missing frame so that the particle's motion appears smooth. To find out where to put the particle in the missing frame, you can do exactly what Isaac and Gottfried did to find Zeno's arrow's position at time t_0, namely pick a point that you think is correct and make sure that the positions on the nearby frames are getting arbitrarily close to that point.

This process is a little complicated, so let's abstract and simplify a little. Notice that, although the arrow was moving in a 3-dimensional world, we could use a single number, its distance to the target, to represent its position. So for each instant of time before 1:00 PM, we have a number representing its position. If we make a list out of these numbers for its positions at 1 minute before 1:00 PM, $\frac{1}{2}$ a minute before 1:00 PM, $\frac{1}{4}$ of a minute before 1:00 PM, and so forth, then these numbers are "approaching" a particular number, which we will call ℓ. Then we are asserting that Zeno's arrow must be at position ℓ at 1:00 PM.

$$(100, 25, 6.25, 1.56, \ldots) = \left(100, \frac{100}{4}, \frac{100}{16}, \frac{100}{64}, \ldots\right)$$
$$= \left(\frac{100}{4^{n-1}} \middle| n \in \mathbb{N}\right) \to \ell = 0$$

It will take us quite a bit of work to turn this notion of "approaching" into a precise definition of *convergence*, but the intuition is clear. This chapter emphasizes and illustrates the important role of the process of starting with an appealing, intuitive idea and then mercilessly refining and specifying the notion until we have finally formulated rigorous definitions that unequivocally capture our generative intuition. The value and power of that process is enormous.

4.2 Convergence

Pinning down the idea of convergence required mathematicians more than 150 years. The challenge is to describe what it means for an infinite *sequence* of numbers to *converge* to a single number, called the *limit*. The intuitive idea is that we have a list of numbers whose values are getting closer and closer to a fixed number. Of course, we first need a precise definition of the objects we'll be studying.

Definition 4.1. A **sequence** of real numbers is an ordered list of real numbers indexed by the natural numbers. So a sequence has a first element a_1, a second element a_2, a third element a_3, and so on. We can denote a sequence in the following ways:

$$(a_1, a_2, a_3, \ldots) = (a_n | n \in \mathbb{N}) = (a_n)_{n \in \mathbb{N}}.$$

We are interested in defining and understanding what it means for a sequence to "converge" to a fixed number ℓ, which intuitively captures the idea that the numbers in the sequence become arbitrarily close to ℓ. Instead of trying to define exactly what it means for a sequence to converge to a number ℓ, let's start by observing some things that had better be true about any definition that captures the notion of "converging".

Observation 1: If the sequence $S = (a_n | n \in \mathbb{N})$ "converges" to a number ℓ, then "eventually" the terms of S had better be "very close" to the number ℓ.

There are two major parts to this observation that we must investigate. What is the precise meaning of "eventually", and what is the best definition of "very close"? We start by trying to make the notion of "very close" more precise, but in the end, the precise definitions of these two ideas will depend on each other. We'll begin by exploring the idea of distance between numbers.

Definition 4.2. Let a and b be two real numbers. Then the **distance** between a and b is defined as $\|b - a\|$, the absolute value of $b - a$. If $\|b - a\| < \varepsilon$, then we say that a and b are **within a distance of ε from each other**.

Recall the definition of the absolute value function: if a is a nonnegative real number then $\|a\| = a$, and if a is a negative real number then $\|a\| = -a$. It follows that, for any real number a, $0 \leq \|a\|$. Also, $\|a\| = 0$ if and only if $a = 0$. We will be using the absolute value function in almost every proof in this chapter, so we should warm up

4.2. Convergence

with a few basic properties. As always, carefully use the definition of the absolute value function rather than a preconceived notion of how it works.

Lemma 4.1. Let $a, b \in \mathbb{R}$.

1. $\|ab\| = \|a\|\|b\|$
2. $a \leq \|a\|$
3. $\|-a\| = \|a\|$

For the absolute value of the difference between two real numbers to be a good notion of distance, it should satisfy one additional property, called the *Triangle Inequality*. Essentially, this inequality captures the notion that it is always shorter to go directly from point a to point b than it is to go first from a to c then from c to b.

Triangle Inequality 4.2. Let a, b, c be real numbers. Then

$$\|b - a\| \leq \|b - c\| + \|c - a\|.$$

(Hint: Note that $\|b - a\|^2 = (b - a)^2$.)

The triangle inequality has several useful, equivalent formulations. Here is one.

Corollary 4.3. Let $a, b \in \mathbb{R}$. Then $\|a\| - \|b\| \leq \|a+b\| \leq \|a\| + \|b\|$.

Recall that we are attempting to make precise some of the vague terms in the following observation:

Observation 1: If the sequence $S = (a_n | n \in \mathbb{N})$ "converges" to a number ℓ, then "eventually" the terms of S had better be "very close" to the number ℓ.

We now have the language to talk about the distance between two numbers, but what does it mean for two numbers to be "very close" to each other? The answer is subtle because "very close" is a relative term. Let's say that the distance between your home and the grocery store is pretty small; perhaps you can drive there in about 11 minutes. But to an ant, the distance is enormous; an ant could probably walk for days before reaching the grocery store from your house. Moreover, when your car is in the shop and you have to walk home from the grocery store carrying groceries in the heat, that distance seems insurmountably large. The point is, for any two distinct points, there is a perspective in which they appear quite far apart.

So, to say that two points are "very close", we must first choose a perspective, which means that we must set an allowable threshold for points to be "very close". If we decide that two points are "very close" when a person can drive between them in under 15 minutes, then your house and the grocery store are "very close". However, if we decide that two points are "very close" only if an ant can walk between them in less than a day, then your house and the grocery store are not "very close". Similarly, if we say that two real numbers within a distance of 0.5 from each other are "very close" to each other, then 2 and 2.1 are "very close" to each other. But if we have stricter standards and require the numbers to be within a distance of 0.001 from each other to be "very close", then 2 and 2.1 are not "very close" to each other.

These examples suggest that we pin down the definition of "very close" by permanently fixing a specific distance for the cut-off. Now that we have articulated a working definition, we need to hold it up against the examples that we know and can build. This process is like the activity of testing a scientific hypothesis through experimentation. Sadly, this first experiment comes back negatively; permanently fixing any specific distance to be the cut-off for "very close" does not produce a reasonable notion of "converging", as you will show in the following exercise.

Exercise 4.4. Suppose that we decreed two real numbers to be "very close" to each other if the distance between them is less than 0.1. Describe a specific sequence $S = (a_n | n \in \mathbb{N})$ whose terms are all "very close" to 5 but whose terms do not "converge" to 5. Of course, we have not yet defined "converge" exactly. Here we want you simply to explain what property your sequence has or does not have that is contrary to your intuitive notion of a sequence "converging" to 5.

So we see that we cannot fix a specific distance as "very close" before defining the notion of "convergence". Instead, for a sequence S to "converge" to ℓ, its terms must be "very close" to ℓ, regardless

4.2. Convergence

of which notion of "very close" is chosen. So the next logical thing to try is to require the sequence's terms to be "very close" for *all* choices of "very close".

***Exercise* 4.5.** Let $S = (a_n | n \in \mathbb{N})$ be a sequence. Suppose all terms in S are "very close" to the number ℓ, regardless of which distance is chosen as the cut-off for "very close". Describe S fully by finding out the exact value of each a_n. Also, explain why no other answer is possible.

The previous exercise shows that requiring every term to be "very close" for every possible distance that we might use as a threshold for determining the meaning of "very close" is far too restrictive. So let's return to the sequence of positions of Zeno's arrow for inspiration:

$$ S = \left(100, \frac{100}{4}, \frac{100}{16}, \ldots\right) = \left(\frac{100}{4^{n-1}} \bigg| n \in \mathbb{N}\right). $$

We began this long discussion to try to formalize the intuitive notion that this sequence "converges" to 0, which corresponds to the distance between arrow and target decreasing to 0. We have not yet found a definition of "converges" that satisfactorily describes even our motivating example, so we must keep looking to find an appropriate definition.

If we require that all terms of the sequence be "very close" to 0 for every possible distance that we might use as a threshold for "very close", then we are in trouble, since any choice of "very close" less than 100 causes a problem with the first term ($\|100 - 0\| = 100$). But starting far away and approaching a value was not a problem for our intuition. Recall that Observation 1 does not require every term in the sequence to be "very close" to ℓ, it just asks that the terms "eventually" get "very close". By including the notion of "eventually" we will be able to find a better notion of "converging".

The intuitive notion of "converging" has to do with the numbers at the end of a sequence rather than with all the numbers of the sequence. We can ignore any finite number of terms at the beginning. For example, consider the following sequence, which simply took our original motivating sequence S and added a few unrelated terms at the beginning:

$$S' = \left(17, 213, 15, 3, 100, \frac{100}{4}, \frac{100}{16}, \ldots\right).$$

This sequence S' also converges to 0, since the tail of this sequence gets very close to 0, just as the tail of S did. This example also foreshadows the idea that a sequence's terms do not need to constantly get closer to ℓ for our intuition to feel that the sequence "converges" to ℓ.

Let's pin down the idea of a tail of a sequence.

Definition 4.3. Let $S = (a_1, a_2, a_3, \ldots)$ be a sequence. Then a **tail** of the sequence S is a set of the form $\{a_m \mid m \geq M\}$, where M is some fixed natural number. Note that S has a different tail for each choice of M, so it does not make sense to talk about the tail.

The key idea is that every property that must "eventually" be true of a sequence can be stated in terms of tails of that sequence. In particular, if a sequence S "converges" to a number ℓ, then for any choice of "very close" there is a tail of the sequence S such that all the terms in that tail are "very close" to ℓ.

Using the new notions of distance and tails, we can reformulate our sense of what it should mean for a sequence to converge to a number ℓ quite precisely.

Observation 1': If a sequence $S = (a_n \mid n \in \mathbb{N})$ "converges" to a number ℓ, then for any cut-off for "very close", there is a tail of S such that all terms in the tail are "very close" to ℓ.

This criterion for convergence is a very technical idea, so let's do a few computational exercises to get familiar with it.

Exercise 4.6. For each sequence below, you will be given a positive real number ε representing the cut-off for "very close". Find a natural number M_ε such that all the terms beyond the M_ε^{th} term lie within the prescribed distance, ε, from ℓ. It is not necessary to find the smallest value for M_ε for which this condition is true. As always, justify your answers.

4.2. Convergence

1. Consider the sequence

$$\left(1, \frac{1}{2}, \frac{1}{3}, \frac{1}{4}, \frac{1}{5}, \ldots\right) = \left(a_n = \frac{1}{n} \mid n \in \mathbb{N}\right) = \left(\frac{1}{n}\right)_{n \in \mathbb{N}}.$$

Find a natural number $M_{0.03}$ such that, for every natural number $k \geq M_{0.03}$, each term a_k lies within a distance of $\varepsilon = 0.03$ from $\ell = 0$.

2. Consider the sequence $(1 - e^{-n} \mid n \in \mathbb{N})$. Find a natural number $M_{0.001}$ such that, for every natural number $k \geq M_{0.001}$, each term $a_k = 1 - e^{-k}$ lies within a distance of $\varepsilon = 0.001$ from $\ell = 1$.

3. Consider the sequence

$$\left(\frac{(-1)^n}{n^2} \mid n \in \mathbb{N}\right).$$

Find a natural number $M_{0.0001}$ such that, for every natural number $k \geq M_{0.0001}$, each term

$$a_k = \frac{(-1)^k}{k^2}$$

lies within a distance $\varepsilon = 0.0001$ from $\ell = 0$.

Finally we are in a position to give a complete definition of convergence of a sequence.

Definition 4.4. A sequence $S = (a_n \mid n \in \mathbb{N})$ **converges** to a number ℓ if and only if for each $\varepsilon > 0$, there exists an $M_\varepsilon \in \mathbb{N}$ such that for any $k \geq M_\varepsilon$, $\|a_k - \ell\| < \varepsilon$. A sequence $T = (b_n \mid n \in \mathbb{N})$ **converges** if there exists a real number ℓ such that the sequence T converges to ℓ.

This definition of convergence of a sequence is so complicated that it requires some real work to understand why each feature of the definition is necessary. The following exercise is basically a copy of a great challenge devised by Carol Schumacher and appearing in her excellent book *Closer and Closer: Introducing Real Analysis*. This exercise asks you to experiment with some inadequate "definitions" of convergence and explain why they aren't correct.

***Exercise* 4.7.** Each of the following statements is an attempt at defining convergence of a sequence. For each statement, explain why that definition would or would not be a good definition of convergence. For each part where you claim that the definition is flawed, include an example of a sequence and a number ℓ that demonstrates why the definition would not be a good definition for convergence.

1. A sequence $S = (a_n | n \in \mathbb{N})$ **converges** to a number ℓ if and only if for each $\varepsilon > 0$, there exists a $k \in \mathbb{N}$ such that $\|a_k - \ell\| < \varepsilon$.

2. A sequence $S = (a_n | n \in \mathbb{N})$ **converges** to a number ℓ if and only if for each $\varepsilon > 0$, there exists an $M_\varepsilon \in \mathbb{N}$ such that for some $k \geq M_\varepsilon$, $\|a_k - \ell\| < \varepsilon$.

3. A sequence $S = (a_n | n \in \mathbb{N})$ **converges** to a number ℓ if and only if for every $M \in \mathbb{N}$ there exists an $\varepsilon > 0$, such that for any $k \geq M$, $\|a_k - \ell\| < \varepsilon$.

4. A sequence $S = (a_n | n \in \mathbb{N})$ **converges** to a number ℓ if and only if for each $M \in \mathbb{N}$ and each $\varepsilon > 0$, there exists a $k \geq M$, such that $\|a_k - \ell\| < \varepsilon$.

5. A sequence $S = (a_n | n \in \mathbb{N})$ **converges** to a number ℓ if and only if for each $\varepsilon > 0$, there exists an $M_\varepsilon \in \mathbb{N}$ such that for any $i \geq j \geq M_\varepsilon$, $\|a_j - \ell\| < \|a_i - \ell\| < \varepsilon$.

6. A sequence $S = (a_n | n \in \mathbb{N})$ **converges** to some number ℓ if and only if for each $\varepsilon > 0$, there exists an $M_\varepsilon \in \mathbb{N}$ such that for any $i, j \geq M_\varepsilon$, $\|a_i - a_j\| < \varepsilon$.

You have now explored at length the reasons for each of the parts of the correct definition of convergence of a sequence. That definition of convergence of a sequence is so complicated that it requires real thought to correctly write its negation. That is your job in the next exercise.

***Exercise* 4.8.** Write out the precise meaning of the following sentence without using the word "not." *The sequence $S = (a_n | n \in \mathbb{N})$ does not converge to ℓ.*

4.2. Convergence

In the previous exercise, you articulated what it means for a sequence not to converge to a particular number ℓ. Often you will be more interested in saying that a sequence doesn't converge to *any* number, that is, that the sequence doesn't converge. The next exercise asks you to write out what conditions will tell you that a sequence does not converge.

Exercise 4.9. Write out the precise meaning of the following sentence without using the word "not." *The sequence $S = (a_n | n \in \mathbb{N})$ does not converge to any number ℓ.* Equivalently, write out the precise meaning of the following sentence without using the word "not." *The sequence $S = (a_n | n \in \mathbb{N})$ does not converge.*

Up to this point, most of our sequences have been written as lists generated by formulas. Some people have strong intuition about such algebraic objects. Others have much more visual and geometric intuition, so we should find a good way to draw a sequence so that we can use geometric intuition. The terms in a sequence are real numbers, and we already have a good way to draw the real numbers: a line. So let's represent the sequence on the real line.

For example, consider the sequence $(a_n = \frac{1}{n} | n \in \mathbb{N})$. All of the terms in this sequence are between 0 and 1, so we should make sure to draw that part of the line very large. Then put a mark for each term in the sequence, and label it as follows.

Of course, we will not be able to draw the infinite number of terms in the sequence, just a representative sample of the first several terms. Knowing how many terms it takes to capture what the sequence is doing is a matter of experience.

We are trying here to use pictures to decide if a sequence converges to a particular number ℓ, so we should include ℓ in the drawing as well. Also, given a distance ε that is the cut-off for how close to ℓ a tail of our sequence must lie, we would like to be able to draw the set of points that are within ε of ℓ. Fortunately, the set of numbers that are within ε of ℓ is an easy set to draw, called an *open interval*.

Definition 4.5. An **open interval** is a subset of the real numbers of the form $\{r \in \mathbb{R} | a < r < b\}$ for two real numbers $a < b$ and is called the "open interval from a to b" or the "open interval with endpoints a and b". We will usually write (a, b) for the open interval from a to b, and call a and b the **endpoints** of the interval.

Note that the symbol "(a, b)" could be an interval or an ordered pair, but the context should make it clear which is intended. And, as we were claiming above, the set of points within distance ε of ℓ is an open interval:

$$\{x \in \mathbb{R}|\ \|x - \ell\| < \varepsilon\} = \{x \in \mathbb{R} | \ell - \varepsilon < x < \ell + \varepsilon\} = (\ell - \varepsilon, \ell + \varepsilon).$$

In other words, the set of numbers that are within distance ε of ℓ is a line segment centered at ℓ, not including the segment's endpoints. Usually, we draw this interval by putting parentheses on the number line at the endpoints and sometimes by shading the segment. For example, if we wanted to see if the sequence $(a_n = \frac{1}{n} | n \in \mathbb{N})$ above has a tail within a distance of 0.1 from $\ell = 0.75$, we would add the following to the drawing.

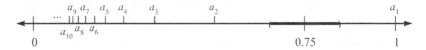

This drawing indicates that no terms of the sequence lie in the interval around 0.75; in particular, no tail lies in that interval. Finding an interval around a number ℓ that contains no tail of the sequence is the same as finding an ε for which the "some tail lies within ε of ℓ" condition for convergence fails. So we can use drawings of sequences to figure out a good choice of ε to use in our proofs of non-convergence.

Exercise 4.10. 1. Consider the sequence

$$A = (a_n = (-1)^n | n \in \mathbb{N}).$$

Show that A does not converge to 1 by finding a specific positive real number ε such that no tail of A lies within a distance ε from 1. Similarly, show that A does not converge to $-1, 0, 2,$ or -2.

4.2. Convergence

2. Consider the sequence $B = (b_n = 2^n | n \in \mathbb{N})$. Show that B does not converge to any number ℓ, that is, show that B does not converge.

3*. Consider the sequence $C = (c_n = \sin(n) | n \in \mathbb{N})$. Show that C does not converge. You may need to look up the precise definition of the trigonometric functions to resolve this challenge thoroughly, but even without that help you should be able to find an appropriate ε by drawing a picture.

It is perhaps useful to think of convergence in terms of the sequence being able to meet any challenge ε. If any challenge ε is proposed, then after some finite number of terms in the sequence are ignored, the remaining tail of the sequence lies within that challenge ε distance of the limit.

Understanding the definition of convergence is tricky because the definition involves infinitely many conditions, namely a condition for each ε bigger than 0. For example, if a sequence converges to 3, we know that after some point in the sequence, all the terms lie within a distance of 1 from 3, namely in the interval $(2, 4)$; perhaps all the terms after the first hundred terms do so. But we also know that eventually all the terms will lie within a distance of 0.1 from 3, namely in the interval $(2.9, 3.1)$; perhaps all the terms after the first million terms do so. We also know that eventually all the terms in the sequence lie within a distance of 0.001 from 3, namely in the interval $(2.999, 3.001)$; perhaps all the terms after the first trillion terms do so. To converge, infinitely many such statements must be true.

To develop some intuition about convergent sequences, let's first look at the examples in the previous exercises and establish which ones converge and which ones do not.

Exercise 4.11. 1. Show that the sequence $\left(\frac{1}{n} \middle| n \in \mathbb{N}\right)$ converges to 0.

2. Show that the sequence $(1 - e^{-n} | n \in \mathbb{N})$ converges to 1.

3. Show that the sequence
$$\left(\frac{(-1)^n}{n^2} \middle| n \in \mathbb{N}\right)$$
converges to 0.

4. Show that the sequence $((-1)^n \mid n \in \mathbb{N})$ does not converge (to any number).

5. Show that the sequence $(2^n \mid n \in \mathbb{N})$ does not converge.

6*. Show that the sequence $(\sin(n) \mid n \in \mathbb{N})$ does not converge.

A convergent sequence can converge to only one number. This uniqueness was not part of the definition of a sequence converging, but it does follow from that definition.

Theorem 4.12. If the sequence $(a_n \mid n \in \mathbb{N})$ converges, then it converges to a unique number.

This last theorem tells us that a convergent sequence approaches exactly one number, which we will call the *limit* of the sequence.

Definition 4.6. If the sequence $(a_n \mid n \in \mathbb{N})$ converges to ℓ, then we say that ℓ is the **limit** of the sequence. In this situation we write

$$(a_n \mid n \in \mathbb{N}) \to \ell.$$

All of this discussion of convergent sequences was motivated by the natural way in which we want to predict the position of Zeno's arrow by looking at its positions at nearby times.

Exercise 4.13. Consider the sequence

$$S = \left(100, \frac{100}{4}, \frac{100}{16}, \ldots\right) = \left(\frac{100}{4^{n-1}} \mid n \in \mathbb{N}\right).$$

Check that S converges to 0, as referees Isaac and Gottfried claimed.

One of the most common examples of convergence arises when we think about decimal numbers. Every decimal number is the limit of a sequence of rational numbers, as you will prove in the next theorem.

Theorem 4.14. Every real number is the limit of a sequence whose terms are all rational numbers. (Hint: How can you tell if a decimal number is rational?)

In a fundamental sense, when we write a decimal number, we are implicitly using the idea of the limit of a sequence. So you have really known about convergent sequences since elementary school days.

4.2. Convergence

Thus far we have represented sequences as lists, as formulas, and as a bunch of marks on the real line. The number line representation has the most obvious geometric uses, but it was messy. Perhaps we can find another graphical representation that doesn't have this problem. Fortunately, we all learned such a technique years ago: graphing. We're used to graphing functions like $f(x) = 3\sqrt{x+5}$, whose domain and codomain are subsets of $\mathbb{R}$. We can view a sequence as a function from $\mathbb{N}$ to $\mathbb{R}$, because for each natural number n, the sequence gives us a real number, namely, the nth number in the sequence.

Consider the sequence $S = \left(a_n = 1 + (-\frac{3}{4})^n \mid n \in \mathbb{N}\right)$; we could graph S as follows.

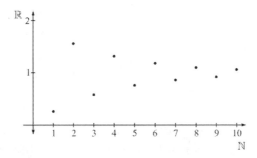

This sequence clearly will converge to $\ell = 1$, but what does that mean in terms of this graphical representation? Well, rather than being a point in one copy of $\mathbb{R}$, ℓ is a horizontal line in this new picture. And eventually being within ε from ℓ means that to the right of some point, all points on the graph are inside an ε-tube of the line representing ℓ. For example, this sequence is eventually within 0.2 from $\ell = 1$.

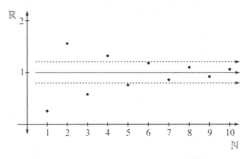

When drawing this picture, we are thinking of a sequence S as a function from $\mathbb{N}$ to $\mathbb{R}$. So this new representation of S is just the graph of this function.

Now let us use this new representation method to investigate our motivating example. We start by drawing the representation of $S = \left(\frac{100}{4^{n-1}} \mid n \in \mathbb{N}\right)$ and including the limit.

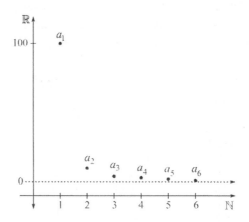

A sequence S converging to a limit ℓ corresponds to the graph of the sequence approaching the horizontal asymptote $y = \ell$.

4.3 Existence of Limits

Thus far, every convergent sequence that we've considered has had an obvious limit. The sequence $\left(\frac{100}{4^{n-1}} \mid n \in \mathbb{N}\right)$ obviously converged to $\ell = 0$; the sequence $(1 - e^{-n} \mid n \in \mathbb{N})$ obviously converged to $\ell = 1$. In this section, we will explore conditions under which we can prove that a sequence does converge even when we may not be able to state explicitly what the limit is.

Consider the sequence

$$S = \left(1, 1 - \tfrac{1}{2}, 1 - \tfrac{1}{2} + \tfrac{1}{4}, 1 - \tfrac{1}{2} + \tfrac{1}{4} - \tfrac{1}{8}, 1 - \tfrac{1}{2} + \tfrac{1}{4} - \tfrac{1}{8} + \tfrac{1}{16}, \ldots\right)$$

$$= \left(\sum_{k=1}^{n} \frac{1}{(-2)^{k-1}} \mid n \in \mathbb{N}\right) = (1, 0.5, 0.75, 0.625, 0.6875, \ldots)$$

It may not be completely obvious from looking at the numbers that

4.3. Existence of Limits

this sequence converges, but we can think of the same sequence geometrically as follows. Imagine Zeno standing on the number line at 0. He takes a step 1 unit to the right and then writes down his position. Then he takes a step 0.5 units to the left and writes down his position. Then he takes a step 0.25 units to the right and writes down his position. Repeating this process produces the sequence S above. It's intuitively obvious that Zeno's position converges because he is alternately moving left then right and his steps are decreasing to 0.

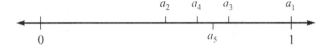

So we believe that this sequence converges, but what is the limit? Well, some of you may remember your calculus really well and may have noticed that S is the sequence of partial sums of a geometric series, so you have a formula that tells you the limit. But being convinced that S converges did not really depend on happening to know a way to figure out the exact limit.

For an example that you really can't easily compute an explicit a limit for, consider the sequence

$$T = \left(1, 1+\tfrac{1}{4}, 1+\tfrac{1}{4}+\tfrac{1}{9}, 1+\tfrac{1}{4}+\tfrac{1}{9}+\tfrac{1}{16}, 1+\tfrac{1}{4}+\tfrac{1}{9}+\tfrac{1}{16}+\tfrac{1}{25}, \ldots\right)$$
$$= \left(\sum_{k=1}^{n} \frac{1}{k^2}\right) = (1, 1.25, 1.36\overline{1}, 1.4236\overline{1}, 1.4636\overline{1}, \ldots),$$

which does converge, but to some number shrouded in mystery.

Let's turn our minds toward looking at properties of sequences with an eye toward finding characteristics of a sequence that will guarantee that it converges even when we can't state its limit.

So let's take a closer look at some of the convergent sequences we've seen

$$S_1 = \left(\frac{100}{4^{n-1}} \,\middle|\, n \in \mathbb{N}\right) = (100, 25, 6.25, 1.57, 0.39, \ldots)$$

$$S_2 = \left(\frac{1}{n} \,\middle|\, n \in \mathbb{N}\right) = \left(1, \frac{1}{2}, \frac{1}{3}, \frac{1}{4}, \frac{1}{5}, \frac{1}{6}, \ldots\right)$$

$$S_3 = (1 - e^{-n} | n \in \mathbb{N}) = (0.632, 0.865, 0.950, 0.982, \ldots)$$

$$S_4 = \left(\frac{(-1)^n}{n^2} \Big| n \in \mathbb{N}\right) = \left(-1, \frac{1}{4}, -\frac{1}{9}, \frac{1}{16}, -\frac{1}{25}, \ldots\right)$$

and compare them to some sequences that don't converge.

$$S_5 = ((-1)^n | n \in \mathbb{N}) = (-1, 1, -1, 1, -1, 1, -1, 1, \ldots)$$

$$S_6 = (2^n | n \in \mathbb{N}) = (2, 4, 8, 16, 32, 64, 128, \ldots)$$

Then we will try to pick out some special features that informed our intuition. Fortunately, we noticed one property long ago: each term in S_1 is smaller than the previous term. Similarly, each term in S_2 is also smaller than its predecessor, and each term in S_3 is larger than the previous term. Let's give these two (related) properties names and precise definitions.

Definition 4.7. A sequence $(a_n | n \in \mathbb{N})$ is called **increasing** if for any $j < k$, $a_j \leq a_k$. The sequence is called **decreasing** if for any $j < k$, $a_j \geq a_k$. A sequence is called **monotonic** if it is either increasing or decreasing. Note that a constant sequence, $S = (c, c, c, \ldots)$, is both increasing and decreasing. If you are drawing a sequence on the real line, then it is monotonic if and only if it only moves in one direction (which includes the possibility of sometimes staying still as well).

A second property of the sequence from Zeno's arrow jumps out as well: the terms in the sequence all lie between two fixed numbers, in the case of the arrow, 0 and 100. Most of these examples are bounded. Having all the terms lie between two fixed numbers sounds useful, so let's give this property a formal definition.

Definitions 4.8. A sequence $(a_n | n \in \mathbb{N})$ is **bounded from above** if there is a real number A such that $a_k \leq A$ for all $k \in \mathbb{N}$. Similarly, the sequence is **bounded from below** if there is a real number B such that $B \leq a_k$ for all $k \in \mathbb{N}$. A sequence is called **bounded** if it is bounded from below and bounded from above; equivalently, a sequence is bounded if there is a real number C such that for all $k \in \mathbb{N}$, $\|a_k\| \leq C$.

4.3. Existence of Limits

Exercise **4.15**. Consider the six sequences above, S_1 through S_6. They are given as formulae and as lists of numbers. Also give the visual representations, using the number line and the 2-dimensional graph. Then make a chart that includes one column for the name of the sequence, one to say if it is monotonic or not, one to say if it is bounded or not, and one to say if it converges or not. Use this chart to make *several* conjectures about the relationships among the conditions of being monotonic, bounded, and convergent. (Not all of your conjectures need to involve all three ideas.)

As we have noted, the sequence S_1 from Zeno's arrow is decreasing and bounded. Since the sequence is decreasing, any limit must be smaller than all of its terms (that is, the limit must be a lower bound), but most lower bounds, for example as -3, are not the limit of the sequence. If we graph this sequence, our intuition would guess the limit to be the value of the horizontal asymptote toward which the values tend. In relationship to S_1, we can describe the number 0 as the *greatest lower bound* for the sequence.

It turns out that the existence of a greatest lower bound for a bounded subset of $\mathbb{R}$ is actually an axiom for the real numbers. (An axiom is a fact that we assume without justification.)

Greatest Lower Bound Axiom. Let $S \neq \emptyset$ be a subset of the real numbers that has a lower bound, that is, there is a real number L such that, for every $s \in S$, $L \leq s$. Then there exists a *greatest lower bound* for S, $\inf(S)$, called the **infimum** of S, with the following properties:

1. If $s \in S$, then $\inf(S) \leq s$.

2. If B is a real number such that $B \leq s$ for all $s \in S$, then $B \leq \inf(S)$.

Similarly, non-empty sets that are bounded above have a least upper bound. This *least upper bound* is called the **supremum**, written $\sup(S)$. The infimum and supremum of a set are unique.

Exercise **4.16**. Write out a careful definition of the Least Upper Bound Axiom.

***Exercise* 4.17.** Check that the infimum of a set is unique, if it exists. Carefully use only the properties guaranteed by the axiom, not your intuitive understanding of what the words should mean.

***Exercise* 4.18.** For each of the following subsets of $\mathbb{R}$, argue whether or not the set has an infimum and supremum. Compute the infima and suprema that exist and justify your computations.

1. $\mathbb{Q}$
2. $(2, 5) \cup \{17\}$
3. $\{\frac{1}{n} | n \in \mathbb{N}\}$

The greatest lower bound axiom allows us to prove that bounded monotone sequences converge.

***Theorem* 4.19.** Bounded monotonic sequences converge.

This theorem is restated below, specifying the limits to which bounded, monotonic sequences converge.

***Theorem* 4.20.** Let $S = (a_n | n \in \mathbb{N})$ be a sequence. If S is increasing and bounded above, then S converges to

$$\ell = \sup\left(\{a_n | n \in \mathbb{N}\}\right).$$

If S is decreasing and bounded below then S converges to

$$\ell = \inf\left(\{a_n | n \in \mathbb{N}\}\right).$$

Monotonicity is not required for convergence, but boundedness is required.

***Theorem* 4.21.** Unbounded sequences do not converge.

This theorem can be phrased more positively as:

***Theorem* 4.22.** Suppose the sequence $S = (a_n | n \in \mathbb{N})$ converges, then S is bounded.

We have now dealt with monotonic sequences and we know which of them converge, namely, those that are bounded. But there are many sequences that we feel intuitively must converge, but that are not

4.3. Existence of Limits

monotonic. For example, the following three sequences all obviously converge to 0.

$$A = \left(1, \frac{1}{2}, \frac{1}{3}, 7, 9, \frac{1}{4}, \frac{1}{5}, \frac{1}{6}, \frac{1}{7}, \frac{1}{8}, \frac{1}{9}, \ldots\right)$$

$$B = \left(1, 0, \frac{1}{2}, 0, \frac{1}{3}, 0, \frac{1}{4}, 0, \frac{1}{5}, 0, \frac{1}{6}, 0, \ldots\right)$$

$$C = \left(1, -1, \frac{1}{2}, -\frac{1}{2}, \frac{1}{3}, -\frac{1}{3}, \frac{1}{4}, -\frac{1}{4}, \ldots\right)$$

None of these sequences is monotonic, and yet convergence seems obvious. This convergence seems obvious because each of these sequences contains, buried inside it, a bounded monotonic sequence that we know converges, and the other terms sort of play along. For convenience, let's give a formal definition of a sequence that uses only some of the terms.

Definition 4.9. Let $S = (s_n | n \in \mathbb{N})$ be a sequence. Then a **subsequence**, T, of S is a sequence obtained from S by omitting some of the terms of S while retaining the order. So T can be written in the form $(t_k = s_{n_k} | k \in \mathbb{N})$ subject to the condition that if $i < j$, then $n_i < n_j$.

This definition is really hard to parse; in particular, the subscript with its own subscript can be bewildering. So let's do an example. Consider the sequence

$$S = (s_n = 1 + 3n | n \in \mathbb{N})$$
$$= (4, 7, 10, 13, 16, 19, 22, 25, 28, 31, \ldots),$$

which has the following two subsequences (among many others):

$$T = (t_k | k \in \mathbb{N}) = (4, 10, 16, 22, 28, 34, \ldots) \quad \text{and,}$$
$$U = (u_k | k \in \mathbb{N}) = (7, 10, 25, 28, 31, 34, 37, \ldots).$$

We could describe T as the sequence containing the odd terms (first, third, fifth...) from S (in the same order). So the first term of T is the first term of S; the second term of T is the third term of S, and similarly, the third term of T is the fifth term of S. In symbols, $t_1 = s_1$, $t_2 = s_3$, $t_3 = s_5$, and so forth. In particular, $t_k = s_{2k-1}$, so $n_k = 2k - 1$. It's easy to check that, if $i < j$, then $n_i = 2i - 1 < 2j - 1 = n_j$.

We could describe the subsequence U as the sequence formed from S by dropping the first and fourth through seventh terms. This subsequence is a much less regular pattern, so it will be harder to find a formula for n_k, but having an easy formula is not required for being a subsequence. In this case, $u_1 = s_2$, $u_2 = s_3$, $u_3 = s_8$, $u_4 = s_9$, $u_5 = s_{10}$, and so on. So in the notation $T = (t_i = s_{n_i})$, the subscripts $n_1 = 2, n_2 = 3, n_3 = 8, n_4 = 9, n_5 = 10$, and so forth.

Although it is important to understand this definition of subsequence, most of the time it will be obvious from the description that the terms of a potential subsequence are in the same order as in the parent sequence.

Exercise 4.23. Let A, B, and C be the sequences from above:

$$A = \left(1, \frac{1}{2}, \frac{1}{3}, 7, 9, \frac{1}{4}, \frac{1}{5}, \frac{1}{6}, \frac{1}{7}, \frac{1}{8}, \frac{1}{9}, \dots\right)$$

$$B = \left(1, 0, \frac{1}{2}, 0, \frac{1}{3}, 0, \frac{1}{4}, 0, \frac{1}{5}, 0, \frac{1}{6}, 0, \dots\right)$$

$$C = \left(1, -1, \frac{1}{2}, -\frac{1}{2}, \frac{1}{3}, -\frac{1}{3}, \frac{1}{4}, -\frac{1}{4}, \dots\right).$$

For each sequence, find a monotonic subsequence and describe n_k explicitly, checking that n_k increases with k.

Omitting terms from a sequence to produce a subsequence moves all subsequent terms towards the beginning of the sequence. Let's state this observation more usefully in the following lemma.

Lemma 4.24. Let $S = (s_n | n \in \mathbb{N})$ be a sequence and $T = (t_k = s_{n_k} | k \in \mathbb{N})$ be a subsequence of S. Then, for every $k \in \mathbb{N}$, $k \le n_k$.

Some properties of parent sequences are not inherited by their subsequences.

Exercise 4.25. 1 Find a sequence that is not bounded but that contains a bounded subsequence.

4.3. Existence of Limits

2 Find a sequence that is not monotonic but that contains a monotonic subsequence.

3 Find a sequence that does not converge but that has a subsequence that does converge.

4 Can you find a single sequence that works for all three parts of this exercise above?

Some other properties of parent sequences are inherited by their subsequences.

***Theorem* 4.26.** Let S be a bounded sequence and S' be a subsequence of S; then S' is bounded.

***Theorem* 4.27.** Let T be a monotonic sequence and T' be a subsequence of T; then T' is monotonic.

***Theorem* 4.28.** If the sequence $S = (a_n | n \in \mathbb{N})$ converges to ℓ and S' is a subsequence of S, then S' converges to ℓ.

This last theorem gives us a strategy to find the limit of some convergent sequences. If the limit of a sequence exists, it is the same for a sequence and its subsequences. So we can use convergent subsequences to propose possible values for the limit. The following is a key technical lemma in this strategy. It is difficult, and you should try to draw several pictures using both visual representations of sequences to help outline your proof.

***Theorem* 4.29.** Every sequence has a monotonic subsequence.

***Corollary* 4.30.** Every bounded sequence has a convergent subsequence.

Having a convergent subsequence is not enough for us to conclude that the whole sequence converges. For the whole sequence to converge, the other terms in the sequence must get close to the values in the subsequence and the limit of that subsequence. Specifically, if a sequence does converge, then every pair of terms in a tail get close to one another.

***Theorem* 4.31.** Let $S = (a_n | n \in \mathbb{N})$ be a sequence that converges to a number ℓ and let $\varepsilon > 0$. Then there exists an $M_\varepsilon \in \mathbb{N}$ such that for any $j, k \geq M_\varepsilon$, $\|a_j - a_k\| < \varepsilon$.

This theorem tells us that in convergent sequences terms eventually get close to one another. That property has a name.

***Definition* 4.10.** A sequence $(a_n | n \in \mathbb{N})$ is called a **Cauchy sequence** if for every $\varepsilon > 0$ there exists an N_ε such that for all $j, k \geq N_\varepsilon$, $\|a_j - a_k\| < \varepsilon$.

Note that this definition does not merely say that the distance between *consecutive* terms is small; it says something much stronger. It says that, after some point, the distance between any two subsequent terms must be small.

***Exercise* 4.32.** Find an example of a sequence such that the distance between consecutive terms decreases to 0 but the sequence does not converge. $1, 1+\frac{1}{2}, 1+\frac{1}{2}+\frac{1}{3},$ etc

The sequence you produced in the previous exercise is not a Cauchy sequence because every tail contains pairs of terms that are far apart.

***Exercise* 4.33.** Write out the precise meaning of the following sentence without using the word "not". *The sequence $S = (a_n | n \in \mathbb{N})$ is not Cauchy*.

***Exercise* 4.34.** Check directly (not as a corollary) that the sequence

$$(3, 2.1, 2.01, 2.001, \ldots) = \left(2 + \left(\frac{1}{10}\right)^{n-1} \bigg| n \in \mathbb{N}\right)$$

is a Cauchy sequence.

Every Cauchy sequence is bounded.

***Theorem* 4.35.** Suppose the sequence $T = (b_n | n \in \mathbb{N})$ is Cauchy; then T is bounded.

***Exercise* 4.36.** In Theorem **4.22** you proved that convergent sequences are bounded. You have now proved that Cauchy sequences are bounded. Compare your proofs of these two facts.

In the theorem before the definition of Cauchy, you actually proved that convergent sequences are Cauchy sequences. For the record, let's state that fact explicitly here.

***Theorem* 4.37.** Let S be a convergent sequence. Then S is a Cauchy sequence.

4.3. Existence of Limits

The property of being Cauchy does not require knowledge of an ellusive ℓ. The definition of being a Cauchy sequence just refers to the terms of the sequence itself. In fact, the Cauchy and convergent properties are equivalent.

***Theorem* 4.38.** Let S be a sequence. Then S is convergent if and only if S is Cauchy.

This last theorem tells us that we can take the definition of a Cauchy sequence as the definition of convergence. In some sense, the Cauchy definition has an advantage over the converging to ℓ definition because the Cauchy condition is intrinsic to the sequence. However, in some cases the definition of convergence that includes the limit ℓ is more useful. Since you have now proved that the two definitions are equivalent, you can appeal to either definition depending on which is convenient for the purpose.

This section is about producing and using the formal definition of convergence. One good way to get better at using the definition of convergence is to conjecture and prove some theorems involving arithmetic combinations of sequences. That investigation is your challenge in the following exercise.

***Exercise* 4.39.** 1. Let $S = (a_n | n \in \mathbb{N})$ be a sequence and let c be a real number. Think of some ways to combine the terms of S with the constant c to create new sequences. Make and prove some conjectures that relate the convergence or non-convergence of S with the convergence or non-convergence of the resulting sequences.

2. Let $S = (a_n | n \in \mathbb{N})$ and $T = (b_n | n \in \mathbb{N})$ be sequences. Think of some ways to combine the terms of these two sequences to create new sequences. Make and prove some conjectures that relate the convergence or non-convergence of S and T with the convergence or non-convergence of the resulting sequences.

In this section, you have explored properties of convergent sequences and investigated conditions on sequences that lead to convergence. For example, you proved that monotonic, bounded sequences converge, and that every convergent sequence is bounded. You proved that subsequences of a convergent sequence converge to the same limit as the parent sequence. You proved that for a sequence, being Cauchy and being convergent are equivalent. You proved theorems about combining sequences arithmetically. All of these results require a detailed understanding of the subtleties of the definition of convergence. Look back at this section and think through the proofs of all these theorems until the details of the definition of convergence and how the definition of convergence is used in proofs become crystal clear to you.

One of the strategies of mathematical inquiry is to pin down an intuitively appealing idea. Making a vague idea specific is like looking through a high-powered microscope. Not only does that idea become crystal clear, but we are then able to see variations and subtleties about it that were invisible at the intuitive level. Your exploration of convergence of sequences demonstrates a powerful technique of distilling and developing potent ideas.

4.4 Continuity

It seems as though Zeno's victory on the archery range is secure. But is the evidence completely ironclad? The reason we might resist awarding Zeno his gold medal right now is that referees Isaac and Gottfried *chose* to measure Zeno's arrow's position at specific instants of time, namely, $\left(\frac{1}{2}\right)^n$ seconds before 1 PM. A careful head judge might find this evidence incomplete. It is conceivable that if the referees had used other instants of time, perhaps $\left(\frac{1}{10}\right)^n$ seconds before 1 PM, then Zeno's arrow might not have appeared to be approaching the bull's eye. So let's rethink our analysis of convergence, but this time imagining that we recorded the positions of the arrow at every single instant before 1 PM.

Recall how we pinned down the idea of convergence of a sequence. We started with a general idea of convergence of a sequence, namely, that a sequence converges to a limit if "eventually" the terms of the sequence become "very close" to the limit. Then we found a way to specify what "eventually" and "very close" really mean. We can use the insights from that analysis to deal with the situation where

4.4. Continuity

we know all the positions of Zeno's arrow before 1 PM rather than just selected moments. Again, the intuitive idea we are trying to capture is that the positions of the arrow are getting closer and closer to a fixed place. Let's begin by noting that the positions of the arrow at every moment before 1 PM are recorded by a function, namely every time before 1 PM gives us a number that is the position of the arrow at that instant. So when we are thinking about convergence or limits in this setting, we are thinking about analyzing functions.

We are interested in defining and understanding what it means for a function that is defined everywhere except x_0 (in the case of the arrow $x_0 = 1$ PM) to converge to a number ℓ or, equivalently, to have a limit ℓ, which intuitively captures the idea that the values of the function become arbitrarily close to ℓ. We are going to undertake the same analysis that we did when we were understanding the idea of convergence of a sequence in the last section. And, in fact, the answer that we arrive at will be extremely similar, so the next couple of pages should seem largely repetitive. (But they are important.)

Instead of trying to define exactly what it means for a function to have a limit ℓ, let's start by observing some things that had better be true about any definition that captures the notion of limit.

Observation 2: If the function $f : \mathbb{R} \to \mathbb{R}$ has a limit ℓ at a point x_0, then "eventually" the values of $f(x)$ had better be "very close" to the number ℓ.

There are two major parts to this observation that we must investigate. What is the precise meaning of "eventually", and what is the best definition of "very close"?

As before when we had this discussion in the context of convergence of a sequence, the answer is subtle because "very close" is a relative term. As before, permanently fixing any specific distance to be the cut-off for "very close" does not produce a reasonable notion of limit.

Just as in the case of convergent sequences, our solution comes from including the notion of "eventually" at the same time that we discuss "very close".

In this case, the intuitive notion of having a limit has to do with the numbers near x_0 rather than with all the values of the function. We can ignore any values of $f(x)$ for x far away from x_0.

Let's pin down the idea of "close to x_0". In the case of sequences, we talked about looking at the values of the sequence in a tail of the sequence. Here we just want to talk about values of the function in a "neighborhood" of x_0. The points in a *neighborhood* of x_0 are just those that lie in a small interval around x_0.

The key idea is that function values as x approaches x_0 are referring to values of the function evaluated at points x that lie in an interval around x_0. In particular, if a function $f(x)$ has a limit ℓ as x approaches x_0, then for any choice of "very close" to ℓ, we want points near x_0 to have function values "very close" to ℓ.

Observation 2′: If a function $f(x)$ has a limit ℓ as x approaches x_0, then for any cut-off for "very close" around ℓ, there is a neighborhood of x_0 such that for any point x in that neighborhood the value of $f(x)$ is "very close" to ℓ.

This criterion for a function approaching a limit is a very technical idea, so let's do a few computational exercises to get familiar with it.

***Exercise* 4.40.** For each function below, you will be given a positive real number ε representing the cut-off for "very close". Find a size $\delta > 0$ such that all the values of the function for points in the δ-neighborhood of x_0 lie within the prescribed distance, ε, from ℓ. It is not necessary to find the largest δ-neighborhood for which this condition is true. As always, justify your answers.

1. Consider the function $f(x) = 5x$. Find a $\delta > 0$ such that, for every x in the δ-neighborhood of 0, each value of $f(x)$ lies within a distance of $\varepsilon = 0.03$ from $\ell = 0$.

2. Consider the function $f(x) = e^x$. Find a $\delta > 0$ such that, for every real number x in the δ-neighborhood of 0, $f(x)$ lies within a distance of $\varepsilon = 0.001$ from $\ell = 1$.

It is helpful to take a graphical look at this issue of convergence of a function to a limit. As in the case of convergence of a sequence, we

some inadequate "definitions" of limit and explain why they aren't correct.

Exercise 4.41. Each of the following statements is an attempt at defining the idea of the limit of a function. For each statement, explain why that definition would or would not be a good definition of limit. For each part where you claim that the definition is flawed, include an example of a function and a number ℓ that demonstrates why the definition would not be a good definition for limit.

1. A function $f : \mathbb{R} \to \mathbb{R}$ has a **limit ℓ at a point** $x_0 \in \mathbb{R}$ if and only if for every $\varepsilon > 0$, there exists a $x \in \mathbb{R}$ with $0 < \|x - x_0\|$ such that $\|f(x) - \ell\| < \varepsilon$.

2. A function $f : \mathbb{R} \to \mathbb{R}$ has a **limit ℓ at a point** $x_0 \in \mathbb{R}$ if and only if for every $\varepsilon > 0$, there exists a $\delta > 0$ such that there exists a $x \in \mathbb{R}$ with $0 < \|x - x_0\| < \delta$ such that $\|f(x) - \ell\| < \varepsilon$.

3. A function $f : \mathbb{R} \to \mathbb{R}$ has a **limit ℓ at a point** $x_0 \in \mathbb{R}$ if and only if for every $\delta > 0$, there exists an $\varepsilon > 0$, such that for every $x \in \mathbb{R}$ with $0 < \|x - x_0\| < \delta$, $\|f(x) - \ell\| < \varepsilon$.

4. A function $f : \mathbb{R} \to \mathbb{R}$ has a **limit ℓ at a point** $x_0 \in \mathbb{R}$ if and only if for every $\varepsilon > 0$ and each $\delta > 0$, for every $x \in \mathbb{R}$ with $0 < \|x - x_0\| < \delta$, $\|f(x) - \ell\| < \varepsilon$.

5. A function $f : \mathbb{R} \to \mathbb{R}$ has a **limit ℓ at a point** $x_0 \in \mathbb{R}$ if and only if for every $\varepsilon > 0$, there exists a $\delta > 0$ such that, for every $x, x' \in \mathbb{R}$ with $0 < \|x - x_0\| < \|x' - x_0\| < \delta$, $\|f(x) - \ell\| < \|f(x') - \ell\| < \varepsilon$.

You have now explored at length the reasons for each of the parts of the correct definition of limit. The definition of limit of a function is so complicated that correctly writing its negation presents a significant challenge. That is your job in the next exercise.

Exercise 4.42. Write the negation of the following statement without using the word "not". *The function* $f : \mathbb{R} \to \mathbb{R}$ *has a limit ℓ at x_0.*

Up to now, we have discussed the idea of a function having a limit. The property of a function having a limit was critical for capturing the

4.4. Continuity

notion that a physical object moved smoothly through space, without teleporting or disappearing temporarily. We pin this property of functions down and name it continuity. A function is continuous at a point just means that the value of the function at that point is predictable from its neighboring values. Predictable means that the limit exists and that the function value is what is expected, namely, that limit.

Definition 4.12. A function $f : \mathbb{R} \to \mathbb{R}$ is **continuous at a point** $x_0 \in \mathbb{R}$ if and only if for every $\varepsilon > 0$, there exists a $\delta > 0$ such that, for every $x \in \mathbb{R}$ with $\|x - x_0\| < \delta$, $\|f(x) - f(x_0)\| < \varepsilon$.

The above definition tells us what it means for a function to be continuous at a point. A function is continuous if it is continuous at every point.

Definition 4.13. A function $f : \mathbb{R} \to \mathbb{R}$ is **continuous** if it is continuous at every point.

Note that for each point, $x \in \mathbb{R}$, being continuous at x means that for every $\varepsilon > 0$, there is a δ_x satisfying the inequalities in the definition of continuity at the point x, but these δ_x's may well be different for different points x. In other words, for a given ε, you may have to use a smaller δ for one point than for another.

Continuous means that the values of $f(x)$ are predictable from the neighboring values. Specifically, the function has to have a limit at each x and the value of the function at x is that limit.

Exercise **4.43.** Let $f : \mathbb{R} \to \mathbb{R}$ be the function defined by

$$f(x) = \begin{cases} x & \text{if } x \in \mathbb{Q} \\ 1 & \text{if } x \in \mathbb{R} \setminus \mathbb{Q} \end{cases}.$$

1. Is f continuous? If so why, if not, why not?

2. Is f continuous at any point?

Exercise **4.44.** 1. Let $f : \mathbb{R} \to \mathbb{R}$ be the function $f(x) = 2x + 1$. Prove that f is continuous.

2*. Let $g : \mathbb{R} \to \mathbb{R}$ be the function $g(x) = \cos(x)$. Prove that g is continuous.

Notice that if a function has a graph that jumps, then that function is not continuous. An intuitive idea of continuity is that you can draw the graph of a continuous function without lifting your pencil. This description is not rigorous, so you should not use it in your proofs. But you can use the idea that continuous functions have graphs with holes or breaks to inform your intuition.

Many classes of functions are continuous.

***Theorem* 4.45.** For any real numbers a and b, the function $f : \mathbb{R} \to \mathbb{R}$ defined by $f(x) = ax + b$ is continuous.

***Theorem* 4.46.** The function $f : \mathbb{R} \to \mathbb{R}$ defined by $f(x) = \|x\|$ is continuous.

Theorem* 4.47.** The trigonometric functions $\sin(x)$ and $\cos(x)$ are continuous.

Theorem* 4.48.** The exponential function e^x is continuous.

Combining continuous functions through addition, multiplication, or composition yields continuous functions.

***Theorem* 4.49.** Let $f(x)$ and $g(x)$ be continuous functions from $\mathbb{R}$ to $\mathbb{R}$. Then $(f + g)(x)$, defined as $(f + g)(x) = f(x) + g(x)$, is continuous.

***Theorem* 4.50.** Let $f(x)$ and $g(x)$ be continuous functions from $\mathbb{R}$ to $\mathbb{R}$. Then $(fg)(x)$, defined as $(fg)(x) = f(x)g(x)$, is continuous.

***Theorem* 4.51.** Let $f(x)$ and $g(x)$ be continuous functions from $\mathbb{R}$ to $\mathbb{R}$. Then $(g \circ f)(x)$, defined as $(g \circ f)(x) = g(f(x))$, is continuous.

***Corollary* 4.52.** Any polynomial

$$p(x) = a_n x^n + a_{n-1} x^{n-1} + \cdots + a_1 x^1 + a_0$$

is continuous.

4.4. Continuity

Lemma 4.53. Let $f : \mathbb{R} \to \mathbb{R}$ be continuous at x_0. If $f(x_0) > 0$, then there is a $\delta > 0$ such that, for every $x \in \mathbb{R}$ with $\|x - x_0\| < \delta$, $f(x) > 0$. Moreover, there is a $\delta' > 0$ such that, for every $x \in \mathbb{R}$ with $\|x - x_0\| < \delta'$, $f(x) > \frac{f(x_0)}{2}$.

Theorem 4.54. Let $f : \mathbb{R} \to \mathbb{R}$ be a continuous function that is never 0. Then the function $h : \mathbb{R} \to \mathbb{R}$ defined as $h(x) = \frac{1}{f(x)}$ is also continuous.

These theorems allow us to show that a vast collection of functions are continuous, such as

$$f(x) = \sin(e^x) \tan \frac{x^2 + 3x + 4}{x^2 + 1}.$$

We've said that continuous functions can be thought of as functions whose graphs we can draw without lifting our pencils. The theorem that really captures this sense is the Intermediate Value Theorem, which states that if a continuous function takes on two values, then it must also take on every value in between.

Notice that the Intermediate Value Theorem would be false if there were any holes in $\mathbb{R}$. The Greatest Lower Bound Axiom tells us that there are no holes in $\mathbb{R}$, so we should expect to use this axiom in the following proof.

This paragraph gives a significant hint toward the proof of the Intermediate Value Theorem, so if you would like to work on it before reading this hint, just skip to the statement of the theorem below. One proof of the Intermediate Value Theorem uses the Least Upper Bound Axiom as a way of describing a number whose functional value is the one you seek. For example, if you look at the set of all the numbers in the domain whose function values are too low, what could you say about the function's value at the Least Upper Bound (or, if appropriate, the Greatest Lower Bound) of that set?

Intermediate Value Theorem 4.55. Let a and b with $a < b$ be two real numbers and $f : [a, b] \to \mathbb{R}$ be a continuous function. Then for any real number r between $f(a)$ and $f(b)$, there is a real number $c \in [a, b]$ such that $f(c) = r$.

Continuous functions attain all intermediate values between any two values they reach. In addition, any continuous function on a closed interval must have a maximum value and a minimum value. Once again, you might consider using the Greatest Lower Bound Axiom as you strive to locate such maxima and minima.

Theorem 4.56. Let a and b with $a < b$ be two real numbers and $f : [a, b] \to \mathbb{R}$ be a continuous function. Then there is a real number $M \in [a, b]$ such that for every $x \in [a, b]$, $f(x) \leq f(M)$, and there is a real number $m \in [a, b]$ such that for every $x \in [a, b]$, $f(x) \geq f(m)$.

The previous theorem has the necessary hypothesis that the continuous function has a closed interval as its domain. A continuous function whose domain is an open interval or the whole real line may not actually reach a maximum or minimum value, as you will demonstrate in the next exercise.

Exercise 4.57. 1. Find a continuous function $f : \mathbb{R} \to \mathbb{R}$ such that for every real x_0, there is another real number x such that $f(x) > f(x_0)$.

2. Find a continuous function $f : (0, 1) \to \mathbb{R}$ such that for every real $x_0 \in (0, 1)$, there is another real number $x \in (0, 1)$ such that $f(x) > f(x_0)$.

3. Can you find bounded continuous functions for each of parts 1. and 2. above?

Continuity captures one of the basic features we know about objects moving about in the world, namely, that their position at any moment is predictable from their positions at times immediately before and after the time in question. If we are being very careful, we realize that we only asked about the position of Zeno's arrow *before* 1 PM. We could develop a more subtle notion of continuity that only requires the values of a function to the left or right of x_0 to be good predictors of the function's value at x_0.

Again your exploration of continuity illustrated the importance of pinning down an intuitively appealing idea by formulating precise definitions and exploring the consequences. Another basic feature about moving objects is their speed, so we turn our attention to the goal of understanding instantaneous velocity in the next section.

4.5 Zeno's Paradox™

After winning the Gold Medal for archery, Zeno, like many retired athletes, needed to find a new line of work. Zeno turned to speeding. Speeding tickets are the bane of existence for people who speed. So when Zeno conceived of his patented Paradox™ FuzzBuster, he was pretty sure he was about to retire to the lap of luxury. The great advantage of the Paradox™ product over those other radar detecting devices was that it did not involve slowing down! It all would have worked perfectly except that the cops who pulled him over during the test run were officers Isaac and Gottfried, who worked as police officers when not judging at the Olympics. Isaac and Gottfried's mathematical insights would wrest order from the jaws of vehicular anarchy caused by Zeno. But we have gotten ahead of ourselves; the story begins with Zeno putting his new Mustang through its paces.

One spring afternoon our "hero", Zeno, jumped into his Mustang convertible and galloped down the straight springtime highway. This highway was extremely well marked, with mileage markers at every single point along the road. The "30 miles per hour" speed limit signs were a mere blur as Zeno raced by. He kept his speed so that at t minutes after 3 pm he was exactly at the mileage marker t^2. So his position $p(t)$ at time t minutes after 3 pm was $p(t) = t^2$. Soon the serenity of the sunny drive exploded as sirens blared, lights flashed, and the strong arms of the law pulled Zeno over for speeding. Zeno had talked his way out of tons of tickets in his life, and he felt his Paradox™ was easily up to the current challenge. So Zeno had no fear that the approaching officers would overcome his evidence of innocence. But his confidence might have been a bit shaken if he had recognized the two officers approaching his window. The officers walked up to Zeno's rolled down window and asked:

Officer Gottfried: Do you know why I pulled you over, sir?
Zeno: No officer, I don't.

Officer Gottfried: Well, the speed limit is 30 miles per hour, and you were doing 120; that's two miles per minute!

Zeno: Really? When?

Officer Gottfried: At precisely 3:01 pm.

Zeno: You must be mistaken. At 3:01 pm, precisely, I was not moving at all, and I can prove it.

Officer Gottfried: How can you prove it?

Zeno: My Zeno's Paradox™ recorded the whole story. You will see that at precisely 3:01 I was only in one place. Here is an instant photograph supplied by the Paradox™ that shows explicitly where I was at precisely 3:01. The Paradox™ was cleverly located exactly across the street from the 1 mile mileage marker sign. And you see that at that exact moment, the nose of my Mustang was precisely lined up with the 1 mile marker. You see that the picture is time-stamped 3:01 exactly. Can I go now?

Officer Isaac: Hold your horses, Bud. You aren't the only cowboy with a camera. Here is a photo of you at 3:02 exactly with that Mustang nostril lined up at the 4 mile marker. Now if I know my math, and you'd better believe I do, that means you went 3 miles in 1 minute, which is why you are going down my friend.

Zeno: Put those cuffs away. You haven't made your case. The question is not where I was at 3:02, the question is how fast I was going at 3:01 and my snapshot shows I was in one place and I rest my case.

Officer Isaac: You will rest your case alright, and you'll rest it in the slammer, because we've got more evidence. Here's another picture—your Mustang's snozzola at the 1.21 mileage marker at precisely 3:01.1. So you were at the 1 mileage marker at 3:01 and at the 1.21 mileage marker at 3:01.1. So you went 0.21 miles in .1 minutes. That works out to 2.1 miles per minute during that half a minute.

Zeno: I'm getting bored. What does my location at 3:01.1 have to do with the question at hand? We are supposed to be talking about 3:01, and at 3:01 I was in precisely one place.

Officer Gottfried: Unfortunately for you, we had an infinite number of cameras taking an infinite number of pictures. In fact, they took pictures of your positions at every instant around 3:01, and, altogether,

4.5. Zeno's Paradox™

they tell a convincing story about speeding: At 3:01.01 you were at mileage marker 1.0201. That means that you went $1.0201 - 1 = .0201$ miles in 0.01 minutes; that is an average speed of 2.01 miles per minute. At 3:01.001, you were at mileage marker 1.002001. So you went $1.002001 - 1 = 0.002001$ miles in 0.001 minutes. That is an average speed of 2.001 miles per minute. We noticed that your location at each time 3:01 plus h minutes was exactly at mileage marker $p(1 + h) = (1 + h)^2$. So for every interval of time h after 3:01, your average speed during the interval of time from 3:01 until 3:01 $+ h$ was

$$\frac{p(1+h) - p(1)}{h} = \frac{(1+h)^2 - 1}{h}$$

(which equals $2 + h$). You are right that no *one* piece of evidence is conclusive, but the totality of this infinite amount of evidence with arbitrarily small lengths of time tells the story. Your instantaneous velocity at time 3:01 was 2 miles per minute because your average velocities during tiny lengths of time around 3:01 *converge* to 2 miles per minute. Zeno your speeding days are done.

Zeno: Converge? Is that the same "converge" involved in continuity? I admit it looks bad for me and my Paradox™, but I'm not going to give up meekly until you convince me that the concept of *convergence* applies to my speed like it applied to my arrow.

Officer Gottfried: The same concept of convergence that got you a Gold Medal will now get you a speeding ticket.

Officer Isaac: And you'll have plenty of time to ponder this reality during your night in the slammer.

Zeno: Can we speed this up? I've got a germ of an idea about acceleration that I want to work on.

The moral of this story is that speed is not a directly measurable quantity. We can measure the length or the weight or the color of Zeno's car directly. Ignoring relativity, we can measure position with a ruler and we can measure time with a stop watch or clock. But to measure speed, we must measure *other* quantities (time and position) at least twice and compute an *average* speed. As the story above indicates, the closer together our measurements are spaced, the greater accuracy we have about what's going on at a given instant. Cars don't generally drive the same speed for any length of time; even using the cruise control, cars change speeds due to hills and other tiny factors. The solution to the puzzle of making a meaningful statement about instantaneous velocity requires us to make infinitely many average speed computations using pairs of instants of time that get arbitrarily close to zero elapsed time, but zero elapsed time makes no sense for measuring motion. Zeno's *instantaneous velocity* is the number to which the totality of an infinite number of computations of average speeds over progressively shorter intervals of time converges.

Specifically, if Zeno's position on a straight road at every time t is given as any function $p(t)$, then the instantaneous velocity at any specific time t_0 is the number to which the values

$$\frac{p(t_0 + h) - p(t_0)}{h}$$

converge as we select values of h that get close to 0. The instantaneous velocity is the single number that summarizes all the approximations of the speed near time t_0 by taking the limit. In our example, we got a sequence of average velocities computed over progressively shorter intervals of time. By looking at intervals of time of length 1 minute, then 0.1 minutes, then 0.01 minutes, then 0.001 minutes, then 0.0001 minutes (each interval starting at 3:01 PM), we computed the average velocities to get a sequence of average velocities

$$(3, 2.1, 2.01, 2.001, 2.0001, \ldots).$$

We plausibly concluded that this sequence of numbers converges to 2, therefore concluding that Zeno's instantaneous velocity at 3:01 PM was 2 miles per minute.

When putting together the case against Zeno, we considered the position of his car as a function of time. We then used his positions to produce average velocities computed using instants near 3:01 PM. Finally, we computed the limit of these "average velocity" measurements and called this limit his instantaneous velocity.

4.6. Derivatives

When computing this limit, we repeatedly computed the fraction

$$\frac{\text{distance travelled}}{\text{time elapsed}}.$$

This complicated fraction is fundamental in computing average velocities, so we give it a name.

Definition 4.14. Let $f : \mathbb{R} \to \mathbb{R}$ be a function. Let x_0 be an number in the domain of f and define a new function $\Delta(f, x_0) : \mathbb{R} \setminus \{0\} \to \mathbb{R}$ by

$$\Delta(f, x_0)(h) = \frac{f(x_0 + h) - f(x_0)}{(x_0 + h) - x_0} = \frac{f(x_0 + h) - f(x_0)}{h}$$

called the **difference quotient** of f at x_0. Note that this difference quotient is a function of h. When $f(x)$ is the position at time x of a moving car on a straight road, then the numerator is change in position and the denominator h is the elapsed time.

Graphically, the difference quotient of f at x_0 evaluated at h is just the slope of a secant line between two points on the graph of $f(x)$, namely, the two points $(x_0, f(x_0))$ and $(x_0 + h, f(x_0 + h))$.

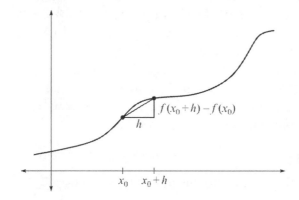

4.6 Derivatives

Let's now return to our hapless "hero," Zeno, who is not doing well in his attempt to avoid his just punishment. Recall that the two officers

Isaac (last name Newton) and Gottfried (last name Leibniz) had presented strong reasoning that Zeno's instantaneous velocity could be computed by taking the limit of

$$\frac{p(t_0 + h) - p(t_0)}{h}$$

as h goes to 0. So let's first give a name to this process of computing the instantaneous velocity.

Definition 4.15. Let $f : \mathbb{R} \to \mathbb{R}$ be a function. For any real number x_0, the **derivative** of f at x_0, denoted $f'(x_0)$, is

$$\lim_{h \to 0} \frac{f(x_0 + h) - f(x_0)}{h},$$

if that limit exists. When the limit does exist, we say that $f(x)$ is **differentiable at** x_0. If $f(x)$ is differentiable at each point x in its domain, then f is **differentiable**.

Differentiable functions can be thought of conceptually as functions whose graphs at each point look straight when looked at under a high-powered microscope. Can you see why the definition of the limit and the definition of derivative tell us that differentiable functions are ones that look straight when magnified? By the way, you can see this effect for yourself using a graphing calculator or a computer. Just graph a differentiable function and then change the scale to have it show you an extremely small interval in the domain and range. You will see that the graph looks like a straight line. The fact that differentiable functions locally look like a straight line means in particular that differentiable functions are continuous.

Theorem 4.58. A differentiable function is continuous.

Not every continuous function is differentiable.

Exercise 4.59. Find a continuous function that has at least one point at which it is not differentiable.

We now proceed essentially to duplicate all our work on continuous functions, but this time considering differentiability instead of

4.6. Derivatives

continuity. Many classes of functions are differentiable. Proving that functions are differentiable is generally more difficult than proving that they are continuous, because we need to prove that the more complicated difference quotient has a limit. The next several theorems allow us to prove that polynomials are differentiable.

When proving that a particular function is differentiable, we must always return to the definition of differentiability, namely, we must prove that
$$\lim_{h \to 0} \frac{f(x_0 + h) - f(x_0)}{h}$$
exists. If $f(x)$ is a differentiable function, then for every fixed value of x, the limit $\lim_{h \to 0} \frac{f(x_0+h)-f(x_0)}{h}$ exists, which means that the limit equals a specific number $f'(x)$. So the derivative of a differentiable function $f(x)$ is another function $f'(x)$. In some cases, we can explicitly write down the function that is the derivative of a given function. Your job in the next theorem is to explain why the derivatives of power functions have the simple form they do. If this theorem proves elusive now, return after you've proved the Product Rule.

Power Rule 4.60. For every natural number n, $f(x) = x^n$ is differentiable and $f'(x) = nx^{n-1}$.

Exercise 4.61. Use the Power Rule to compute Zeno's instantaneous velocity at 3:01 PM.

Theorem 4.62. If $f(x)$ is differentiable and a is a real number, then the function $g(x) = af(x)$ is also differentiable and $g'(x) = af'(x)$.

Theorem 4.63. Let $f(x)$ and $g(x)$ be differentiable functions from $\mathbb{R}$ to $\mathbb{R}$. Then $(f+g)(x)$, defined as $(f+g)(x) = f(x) + g(x)$, is differentiable and $(f+g)'(x) = f'(x) + g'(x)$.

Corollary 4.64. Every polynomial function
$$f(x) = a_n x^n + a_{n-1} x^{n-1} + a_{n-2} x^{n-2} + \cdots + a_2 x^2 + a_1 x + a_0$$
is differentiable and
$$f'(x) = na_n x^{n-1} + (n-1)a_{n-1} x^{n-2} + (n-2)a_{n-2} x^{n-3} \\ + \cdots + 2a_2 x + a_1.$$

Before we get too lackadaisical about how these derivatives are going to proceed, let's point out that products do not work as expected.

Exercise 4.65. Find two differentiable functions $f(x)$ and $g(x)$ for which the derivative of their product is not the product of their derivatives.

Since the derivative of a product is not as simple as one might think, let's analyze what the derivative of the product actually is and why it is so.

Before we proceed with the derivative of the product of two functions, let's introduce an alternative notation for the derivative. If $y = f(x)$ is a function, then its derivative $f'(x)$ can be denoted $\frac{d}{dx}(f(x))$ or $\frac{dy}{dx}$. Notice that this notation reminds us of the definition of the derivative. This notation was carefully designed to do so by one of the inventors of calculus, Gottfried Leibniz. Leibniz thought carefully about the notation so that operations of calculus could be done somewhat mechanically. One of the virtues of calculus is that much calculus work can be done by rote, and Leibniz's carefully crafted notation makes such routine work convenient.

Let's now return to analyzing the derivative of the product of two functions. We will begin by considering the product of two specific, simple functions.

Exercise 4.66. Let $f(x) = ax$ and $g(x) = bx$. What is the derivative of the product, that is, $\frac{d}{dx}(f(x)g(x))$? Of course, you could simply multiply $f(x)$ times $g(x)$ to get the product abx^2 and then take its derivative. That is fine, particularly to check your thinking; however, for this exercise please think about the definition of the derivative and apply the definition to the product directly. That is, consider the quotient:

$$\lim_{h \to 0} \frac{f(x+h)g(x+h) - f(x)g(x)}{h}$$

$$= \lim_{h \to 0} \frac{a(x+h)b(x+h) - (ax)(bx)}{h}$$

$$= \lim_{h \to 0} \frac{(ax+ah)(bx+bh) - (ax)(bx)}{h}$$

4.6. Derivatives

Try to understand the value of that difference quotient by multiplying out the numerator without simplifying. The goal of this exercise is for you to see the relationships among the derivatives of each of the functions, the values of each of the functions, and the derivative of the product.

Recall that differentiable functions look like straight lines locally. So the previous exercise guides us to guess what the derivative of a product should be. Alternatively, if you remember the Product Rule from a calculus course, the following theorem will not be a surprise.

Product Rule 4.67. Let $f(x)$ and $g(x)$ be differentiable functions. Then their product is differentiable and

$$\frac{d}{dx}(f(x)g(x)) = f(x)g'(x) + f'(x)g(x).$$

As long as we are working our way through the combination of functions, we may as well tackle reciprocals and then quotients. Once again, we ask you to analyze a particular function in order to see the relationship among the derivative of the function, its function value, and the derivative of its reciprocal.

Exercise 4.68. Let $f(x) = ax$. Using the definition of derivative, compute the value of $\frac{d}{dx}\left(\frac{1}{f(x)}\right)$. After writing out the definition of the derivative, do some algebraic simplifications with the difference quotient, but think of the constant a as $f'(x)$, so do not cancel a's during your work. The goal of this exercise is for you to think through the definition of the derivative to see how the derivative of the function, the value of the function, and the derivative of the reciprocal are related.

If you were successful with the previous exercise or if you remember the Reciprocal Rule from a calculus course, the following theorem will not be a surprise.

Reciprocal Rule 4.69. Let $f(x)$ be a differentiable function with $f(x_0) \neq 0$. Then

$$\frac{d}{dx}\left(\frac{1}{f(x)}\right) \text{ at } x_0 \text{ equals } -\frac{f'(x_0)}{(f(x_0))^2}.$$

By combining the Reciprocal Rule and the Product Rule, we can formulate the Quotient Rule.

***Quotient Rule* 4.70.** Let $f(x)$ and $g(x)$ be differentiable functions. Then
$$\frac{d}{dx}\left(\frac{f(x)}{g(x)}\right) = \frac{f'(x)g(x) - f(x)g'(x)}{(g(x))^2}$$
for every x for which $g(x) \neq 0$.

To actually take derivatives, the strategy is to individually take the derivatives of some basic functions using the definition of the derivative and then combine those results using rules of combination such as the sum, product, and quotient rules to determine the derivatives of more complicated functions. Let's now turn to trigonometric functions.

The trigonometric functions are differentiable, but again they present a challenge. Each of the basic trigonometric functions has its own difficulties, so let's just start with the sine and cosine.

***Exercise* 4.71.** 1. Write down the difference quotient involved in the limit definition of the derivative of $\sin(x)$ and consider the picture below of the unit circle. Describe the coordinates of A and B in terms of the angles x and $x + h$.

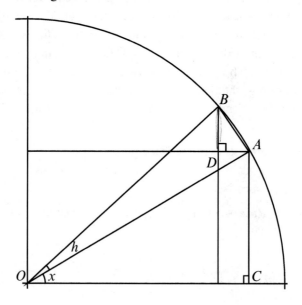

4.6. Derivatives

Use the fact that the radius is basically perpendicular to AB to prove that $\triangle COA$ is similar to $\triangle DBA$. In addition, notice that the length of AB is almost exactly the length of the arc subtended by $\angle AOB$. (In particular, these assertions become closer and closer to true as h goes to 0.) You can now describe the lengths of the sides of both triangles. Use the similarity to compute the difference quotient in another manner.

2. Use the same picture and similar reasoning to deduce the derivative of $\cos(x)$.

The previous exercise correctly suggests why the derivatives of the sine and cosine are what they are; however, pinning down the proofs requires additional analysis.

Theorem* 4.72. *The trigonometric function $\sin(x)$ is differentiable and*
$$\frac{d}{dx}\bigl(\sin(x)\bigr) = \cos(x).$$

Theorem* 4.73. *The trigonometric function $\cos(x)$ is differentiable and*
$$\frac{d}{dx}\bigl(\cos(x)\bigr) = -\sin(x).$$

We can now proceed to the other trigonometric functions by using the reciprocal and quotient rules.

Exercise 4.74. Given the derivatives of $\sin(x)$ and $\cos(x)$, determine the derivatives of the trigonometric functions $\tan(x)$, $\sec(x)$, $\csc(x)$, and $\cot(x)$.

One of the most potent methods for obtaining more complicated functions from simpler ones is to compose functions. Let's see how the derivative of the composition of two functions is related to the derivatives of the two functions involved. Suppose we have two differentiable functions $f(x)$ and $g(x)$ and we consider the composition $g(f(x))$. Let's think about the derivative of the composition, that is, $\frac{d}{dx}(g(f(x)))$. The derivative answers the question, "If we change x by a small amount h, how much will $g(f(x))$ change?". We know that a small change h in x will result in a change of approximately $hf'(x)$ in $f(x)$. And we know that a small change k from the value $f(x)$ will make $g(f(x))$'s value change by about $kg'(f(x))$. So a change of h in x's value will make $f(x)$ change by about $hf'(x)$, which in turn will make g differ by about $hf'(x)g'(f(x))$ from the value $g(f(x))$. So we conclude that the derivative of $g(f(x))$ should be g's derivative at $f(x)$ times the derivative of f at x. Let's do a specific example to illustrate this insight.

Exercise 4.75. Let $f(x) = 2x + 1$ and $g(x) = x^3$. What is the derivative of the composition, that is, $\frac{d}{dx}(g(f(x)))$? Of course, you could simply take the composition, which means to cube $2x + 1$ to realize that $g(f(x)) = (2x+1)^3 = 8x^3 + 12x^2 + 6x + 1$ and then take its derivative. That is fine, particularly to check your thinking; however, for this exercise please think about the definition of the derivative and apply the definition to the composition directly. Try to understand the relationships among the derivatives of each of the functions and the derivative of the composition.

The theorem that captures these insights into the derivative of a composition is called the Chain Rule, as you probably remember from a calculus course.

Chain Rule 4.76. Let $f(x)$ and $g(x)$ be differentiable functions from $\mathbb{R}$ to $\mathbb{R}$. Then $(g \circ f)(x) = g(f(x))$ is differentiable and

$$\frac{d}{dx}\left(g(f(x))\right) = g'(f(x))f'(x).$$

These theorems allow us to take derivatives of a vast collection of continuous functions such as $\sin^3(x)\tan(x^2 + 3x + 4)$.

We have identified a large class of functions that are differentiable. So now let's make a couple of observations about special points at which differentiable functions must have derivative equal to 0. When we think graphically, the derivative of a function is the slope of the tangent line to the graph of that function. The following theorem records that, as expected, the value of a derivative at a local minimum or at a local maximum has derivative equal to 0.

Theorem 4.77. Let $f : \mathbb{R} \to \mathbb{R}$ be a differentiable function. Suppose w is a local maximum of f, that is, there is an open interval (a, b)

containing w such that for every $x \in (a, b)$, $f(x) \leq f(w)$. Then $f'(w) = 0$. Similarly, if w is a local minimum of f, then $f'(w) = 0$.

One theorem that captures the global implications of differentiability is the Mean Value Theorem, which implies that if Zeno drove at a particular average velocity over an interval of time, then at some instant, his instantaneous velocity was that average velocity. This plausible statement can be couched in terms of derivatives.

***Mean Value Theorem* 4.78.** Let $f : [a, b] \to \mathbb{R}$ be a continuous function that is differentiable at each point in (a, b). Then for some real number $c \in (a, b)$,

$$f'(c) = \frac{f(b) - f(a)}{b - a}.$$

***Exercise* 4.79.** Use the Mean Value Theorem to give a new proof that Zeno was speeding sometime between 3:00 PM and 3:02 PM.

Our exploration of the derivative again illustrates the value of pinning down an intuitive idea through making rigorous definitions.

4.7 Speedometer Movie and Position

This discussion of derivatives all emerged from the question of finding instantaneous velocity when we know the position of a car moving on a straight road at each instant. Let's return to moving cars to look at the reverse question, namely, finding the position if we know the instantaneous velocity at each moment.

Avant-garde movies strive for deep meaning, often with no action. These movies are incredibly boring and here we will describe some of the most boring. After their stint as archery referees and on the police force, Newton and Leibniz decided to turn their attention to film. They got in a car, turned the lens on the speedometer, and drove forward on a straight road for an hour. The movie was not edited and presented only the speedometer dial with the needle sometimes moving slowly, sometimes fixed for minutes on end. None of the road could be seen and the action was unrelieved by a glimpse at the odometer. The movie was time-stamped at each moment, so the viewer could see how much of life would be wasted before the merciful conclusion of this "drama". Newton and Leibniz made several of these hour-long movies; however, few people went back to see the sequels.

Since viewers were terminally bored with these movies, Newton and Leibniz decided to pose a question to give their audience something to do. They asked, "How far did the car go during this hour?"

This question turned the movie from a sleeper to a riveting challenge that changed the world.

Exercise **4.80.** Here are some descriptions of the speedometer movies. For each one, figure out how far the car went and develop a method that would work for any such movie.

1. This movie is the most boring of all. For the entire hour-long movie, the speedometer reads 30 mph.

2. This movie has only one change. For the first half hour, the speedometer reads 30 mph, and then instantly changes to read 60 mph for the second half hour.

3. In this movie, the speedometer starts at 0 mph and gradually and uniformly increases by 1 mph each minute to read 60 mph at the end of the hour.

4. In this movie, the speedometer's reading is always t^2 mph where t is the number of minutes into the movie. This car is really moving by the end of the hour, in fact, it may be a rocket ship instead of a car.

5. Now we come to the general case. Suppose you have any such movie in which the speedometer is changing arbitrarily. What strategy could you devise to pin down the distance traveled during the hour to within 1 mile of the actual distance? ... to within 0.1 miles? ...to within .001 miles? ... to pin down the distance exactly?

In answering the previous exercise, you have defined the definite integral. In the following definition, think of the function $f(x)$ as telling the speedometer reading at each time x.

Definition 4.16. Let $f(x)$ be a continuous function on the interval $[a, b]$. Then the **definite integral** of $f(x)$ from a to b is a limit of a

4.8. Applications of the Definite Integral

sequence of approximating values (each approximation being a sum of products) where the nth approximation is obtained by dividing the interval $[a, b]$ into n equal subintervals:

$$[a_0 = a, a_1], [a_1, a_2], [a_2, a_3], \ldots, [a_{n-2}, a_{n-1}], [a_{n-1}, a_n = b],$$

then for each subinterval multiplying its width (which is always $\frac{b-a}{n}$) by the value of the function at its left endpoint (that product would give approximately the distance traveled during that small interval of time) and then adding up all those products. For every choice of n, the number of intervals, we have an approximation of the integral, and the integral is the limit as we choose increasingly large n, which produces increasing small subintervals. In symbols:

$$\int_a^b f(x)dx = \lim_{n \to \infty} \sum_{k=0}^{n-1} f(a_k) \frac{b-a}{n}$$

$$= \lim_{n \to \infty} \sum_{k=0}^{n-1} f\left(a + k \frac{b-a}{n}\right) \frac{b-a}{n}.$$

Leibniz is again responsible for the notation for the integral. Notice that every feature of the notation refers to its definition. The long S shape stands for "sum," the limits of integration tell us over what interval x ranges, the "dx" is the small width and it is next to the $f(x)$, so $f(x)dx$ suggests the distance traveled in the small "dx" interval of time. So adding up those small contributions to the distance traveled gives the total distance traveled.

Yet again, we see how the process of creating a precise definition to specify an intuitively appealing idea has enormous consequences.

4.8 Applications of the Definite Integral

Recall that the derivative was defined to give us meaningful information about the world. If we know the position of a moving car at each instant, the derivative tells us the velocity of the car at each instant because the derivative was defined to accomplish that. That same definition of the derivative can be interpreted as telling us the slope of the graph of a function at each point. By looking at the definition of the derivative, we see that the basic difference quotient involved in the definition can be seen as capturing approximations of the slope of the graph of the function. If we thought about examples in the practical world, we would see other instances where the definition of the derivative turns out to be exactly what we need to describe some aspect of the world. For example, suppose we are trying to understand the buying behavior of people when faced with an enticing product at different potential prices. If the price is set higher, the demand goes down. If the price is lower, the demand goes up. How fast does the demand go down compared to a rise in price? To answer that question one would naturally write down an expression that was, again, a derivative. Derivatives arise so frequently because the basic definition of the derivative captures the basic idea that we often want to know: how does a change in one varying quantity influence a change in another quantity?

The definite integral enjoys a correspondingly broad range of applicability, or perhaps an even greater range of applicability. Very often an effective strategy for computing a desired value is to divide something up into tiny pieces and then add up those contributions. We saw in the last section that the definite integral is exactly what you do to compute the distance traveled if you know the velocity of a moving car at each instant. Often the impulse to divide into tiny pieces and add them up is perfectly captured by the definite integral. So in this section, we want you to become solidly familiar with the definition of the definite integral by demonstrating how the definition arises and is applicable in several situations.

Let's begin by looking at some geometric figures as figures that are growing. We'll start with a square.

***Exercise* 4.81.** Think of a square as growing from a corner such that at each time t, each side has length t. Think of building up the square by adding layers to the two sides opposite a corner.

4.8. Applications of the Definite Integral

What integral would describe the area of a square with side of length s from that perspective?

Next let's consider a cube.

Exercise 4.82. Think of a cube as growing from a corner such that at each time t, each side has length t. Think of building up the cube by adding layers to the three sides opposite a corner. What integral would describe the volume of a cube with side length s from that perspective?

You may have wondered why we chose to have the fixed point in the square and the cube in the previous two exercises be a corner point rather than the point in the center. So let's just return to those same two figures, but look at them as growing from the center.

Exercise 4.83. Think of a square and a cube as growing from the center such that at each time t, each side as length t. Think of building up the square and cube by adding layers to all the sides. What integrals would describe the area and volume of the square and cube with side length s from that perspective?

The definition of the integral arises naturally whenever we can take an object whose area or volume we want to compute and we can think of that object as made of thin layers. Let's think about a pyramid.

Exercise 4.84. Think of a pyramid whose height is the same as each side of its square base. Suppose the height of the pyramid is 100 feet. If you think of that pyramid as consisting of diminishing sized, horizontal squares layered on top of one another from the base to the apex, what integral would describe the volume of the pyramid from that perspective?

Some of the most fun examples of the definite integral arise when we compute the volume of a solid of revolution. Here are a couple of examples.

Exercise 4.85. Write an integral whose value is the volume of the solid created when the region under the graph of the curve $y = x^2$ for $4 \leq x \leq 5$ is revolved about the x-axis.

Exercise 4.86. Write an integral whose value is the volume of the solid created when the region under the graph of the curve $y = x^2$ for $4 \leq x \leq 5$ is revolved about the y-axis.

Many concepts in physics are naturally modeled by the definite integral. Let's consider pressure.

***Exercise* 4.87.** The total water pressure on a side of a swimming pool is the total amount of force pushing against that side. At each point on the wall, the force is measured as pounds per square foot, so if there were uniform pressure of 100 pounds per square foot on a square foot of wall, then the total pressure on that square foot is 100 pounds of force. Unfortunately, pressure varies with depth, so the pressure is not uniform over any square foot of wall since the lower part of that square foot would have a greater pressure per square foot than the upper part. As you will soon see, the integral will come to the rescue. Here's the question:

Suppose a swimming pool has a rectangular wall that is 10 feet wide and 8 feet deep and the water is full to the top of the wall. Suppose the water pressure at depth d feet below the surface of the water is $63d$ pounds per square foot. Write an integral expression whose value is the total pressure on the wall.

Another basic concept in physics is work, which is the product of force and distance. So the integral again is the perfect concept to compute work.

***Exercise* 4.88.** Suppose the force required to hold a spring compressed to a times its natural length ($0 < a < 1$) is $\frac{K}{a}$. Write a definite integral whose value is the total work required to compress the spring from its natural length to half its natural length.

The idea of applying the integral is to find a way to make a stepwise approximation and then realize how that process becomes an integral in the limit. This next application asks you to find the formula for the length of a parametrized curve in the plane.

***Exercise* 4.89.** Suppose a smooth curve in the plane is described by the parametrized functions $(f(t), g(t))$ for $0 \leq t \leq 1$. What integral would describe the length of the curve?

Let's return for a moment to our car moving on a straight road. This time, let's consider acceleration, which measures the rate at which the velocity is changing, that is, the derivative of the velocity.

4.9. Fundamental Theorem of Calculus

***Exercise* 4.90.** Suppose a car drives on a straight road that has mileage markers at each point. Suppose the car begins at a stop at the mileage marker 0. Now suppose at each time t the car accelerates according to a continuous function $a(t)$. What integral expression would tell us where the car is at each time t?

The integral can be used to measure the volume of a mountain.

***Exercise* 4.91.** Suppose the height of a mountain in feet is given by the formula $h(x, y) = x^2 y$ over the rectangular region defined by $400 \leq x \leq 1000$ and $200 \leq y \leq 600$, where x and y are measured in feet. What integral has a value equal to the volume of the mountain over the region described?

These exercises are merely the tip of the iceberg in demonstrating the amazing range of applicability of the definite integral. However, what we have not yet seen is how in the world we could possibly compute the values of those integrals. We know that the values of the integrals would have the significance described in each of the above situations. The next section will show us how we can actually compute a specific answer.

4.9 Fundamental Theorem of Calculus

Since the derivative and the integral really involved the same car moving down the road, there is a clear and natural connection between the two concepts of the derivative and the integral. Namely, there are two ways to look at how far a car traveling along a straight road has traveled. On the one hand, see where the car was at the end and subtract where it was at the beginning in order to compute the net change over that interval of time. The other way is to do the integral procedure. Since both methods yield the same result of the net change in the position of the car, those two methods must produce the same answer. But notice that if we have a position function $p(t)$ that is telling us the position of the car at every time t from some time a to time b, then $p'(t)$ is telling us what the speedometer will be reading at each moment. So we can see that the integral of $p'(t)$ from time a to time b will give the same answer as the difference in the ending position $p(b)$ minus the starting position $p(a)$. This insight, which you will prove next, is the most important insight in calculus and therefore has the exalted title of the *Fundamental Theorem of Calculus*.

Fundamental Theorem of Calculus 4.92. Let $F(x)$ be a function on the interval $[a, b]$ with continuous derivative $F'(x)$. Then

$$\int_a^b F'(x)dx = F(b) - F(a).$$

If you are given a function $g(x)$ and you find a function $h(x)$ such that $h'(x) = g(x)$, then $h(x)$ is called an **anti-derivative** of $g(x)$. So in the Fundamental Theorem of Calculus, the function $F(x)$ is an anti-derivative of $F'(x)$.

The definition of the definite integral tells us that the value of the integral is something meaningful that we want to know, such as the net distance a car has traveled if we are told its velocity at each instant. The Fundamental Theorem of Calculus tells us that to find the value of a definite integral, all we need to do is to find an anti-derivative, plug in two values, and subtract. So the Fundamental Theorem of Calculus is the reason that anti-derivatives are so closely linked with integrals. In fact, we soon start saying "integral" when we mean "anti-derivative."

After we defined the derivative, we proceeded to deduce several theorems that allowed us to compute derivatives of many functions. The Fundamental Theorem of Calculus tells us that if we can find anti-derivatives of functions, then we will be able to compute definite integrals. Computing an anti-derivative requires us to recognize a function as being the result of taking a derivative of another function, its anti-derivative. Therefore, every technique for taking derivatives can be turned into a technique for taking anti-derivatives by looking at the form of the answers after using the derivative method and seeing what function must have been the one whose derivative gave that result. So let's look at various techniques for taking derivatives and, for each one, deduce a corresponding technique of integration, that is, a technique for anti-differentiation.

Let's start with the Power Rule for taking derivatives of functions $f(x) = x^n$. Recall the Power Rule:

Power Rule. For every natural number n, $f(x) = x^n$ is differentiable and $f'(x) = nx^{n-1}$.

4.9. Fundamental Theorem of Calculus

Looking at this theorem in reverse, we have an anti-derivative result. Notice that adding a constant C to a function results in a function with the same derivative.

Anti-derivative Power Rule 4.93. For every natural number n and real number C, $f(x) = nx^{n-1}$ has an anti-derivative $F(x) = x^n + C$. In particular, for every natural number n or $n = 0$ and real number C, $f(x) = x^n$ has an anti-derivative

$$F(x) = \frac{1}{n+1}x^{n+1} + C.$$

This small insight allows us to take anti-derivatives of any polynomial.

Anti-derivatives of Polynomials 4.94. State and prove a theorem that shows how to find an anti-derivative of any polynomial.

We can find anti-derivatives of the basic trigonometric functions, although finding anti-derivatives of other trigonometric functions is a bit trickier.

Trigonometric Anti-derivatives 4.95. State and prove a theorem that shows how to find anti-derivatives of the sine and cosine functions.

Every derivative theorem looked at backwards gives a technique for taking anti-derivatives, so let's see what technique we can deduce from the Chain Rule. Recall the Chain Rule:

Chain Rule. Let $f(x)$ and $g(x)$ be differentiable functions from $\mathbb{R}$ to $\mathbb{R}$. Then $(g \circ f)(x) = g(f(x))$ is differentiable and

$$\frac{d}{dx}\big(g(f(x))\big) = g'(f(x))f'(x).$$

The Chain Rule allows us to recognize certain functions as the result of taking a derivative. Simply stated, if we see a function $h(x)$ that we can recognize as a product $g'(f(x))f'(x)$, then we know that anti-derivatives of that function would be

$$\int h(x)\,dx = g(f(x)) + C.$$

This insight leads to the technique often affectionately referred to as "u-substitution".

***u-Substitution* 4.96.** Give several examples of functions whose antiderivatives you can find by recognizing the functions as the result of an application of the Chain Rule.

The final example we will consider where we look at a derivative rule to deduce an integration technique involves the Product Rule. Recall the Product Rule:

Product Rule. Let $f(x)$ and $g(x)$ be differentiable functions. Then their product is differentiable and

$$\frac{d}{dx}(f(x)g(x)) = f(x)g'(x) + f'(x)g(x).$$

***Integration by Parts* 4.97.** Given two differentiable functions $f(x)$ and $g(x)$, show why an anti-derivative of $f(x)g'(x)$ equals $f(x)g(x)$ minus an anti-derivative of $f(x)g'(x)$. Give several examples of functions whose anti-derivatives you can find by applying this technique of integration by parts. In particular, find an anti-derivative of the logarithm function.

We end this exploration of the integral in a manner analogous to how we concluded our exploration of the derivative. When we explored the derivative, we noticed in the Mean Value Theorem a relationship between the average rate of change of a function over an interval and its derivative at a single point in that interval. Here is the Mean Value Theorem:

Mean Value Theorem. Let $f : [a, b] \to \mathbb{R}$ be a continuous function that is differentiable at each point $x \in (a, b)$. Then for some real number $c \in (a, b)$,

$$f'(c) = \frac{f(b) - f(a)}{b - a}.$$

The analogous theorem for integrals is that the integral over an interval gives the same result as a constant function would give over that interval, if we select the correct value.

***Mean Value Theorem for Integrals* 4.98.** Let $f(x)$ be a continuous function on the interval $[a, b]$. Then for some real number $c \in (a, b)$,

$$\int_a^b f(x)dx = f(c)(b - a).$$

4.10. From Vague to Precise

The definite integral allows us to compute the total distance a car on a straight road will have traveled during an interval of time between time a and time b if we know the velocity $f(x)$ of the car at each time x. The Mean Value Theorem for Integrals assures us that the total distance traveled during the time period from time a to time b could have been accomplished by driving during the whole time at some fixed velocity $f(c)$, where $f(c)$ is a velocity that the car actually did travel at some instant during the journey.

4.10 From Vague to Precise

Our exploration of limits and convergence, continuity, the derivative, and the integral treated the foundational ideas of calculus. Further extensions of these ideas have occupied mathematicians from the time of Newton and Leibniz to the present day. It would be difficult, if not impossible, to find a set of ideas that have had a more profound impact on our ability to understand and describe our world than calculus.

Calculus has applications in virtually every area of study including the physical sciences, biological sciences, business and economics, statistics, and many, many more. One of the applications of most value occurs when calculus is used to optimize some quantity, for example, optimizing profits based on decisions concerning how much to manufacture or optimizing the probability of a decision being a good one in sports or war. Calculus is the language in which descriptions of the universe are couched. Calculus was created to help Newton and Leibniz describe the physical world, but it is now applied in the social sciences as well. There is clearly no limit to the value of calculus as an aid to understanding our world.

Our investigation into calculus went through our process of inquiry (abstraction, exploration, conjecture, justification, application, and extension) four times, investigating sequences, continuity, derivatives, and integrals. The repeated use of these strategies illuminates the cyclical nature of inquiry.

We started by taking a situation in the world and used it as a guide while defining the concepts of convergence and limits in the process called abstraction. We explored those concepts, conjectured and justified patterns, and applied our results to understand the motion of an arrow in space. The insights we developed allowed us to extend our investigation by pinning down the idea of continuous functions. We then again turned to the real world to provide the motivation from which we abstracted the notion of the derivative. Pinning down the definition of the derivative used limits. We saw how to find derivatives of polynomials and some basic trigonometric functions directly from the definition of derivative, and then we saw how functions created by combining other functions via addition, multiplication, and composition let us increase vastly the collection of functions that we could differentiate. Then we again returned to reality to guide us to define the integral. Again we saw how to pin down the definition using limits. We saw how the integral captures things we want to know such as distance traveled or area under curves. Then we saw the connection between the derivative and the integral in the Fundamental Theorem of Calculus.

This life-transforming subject has been developed by using potent strategies of creating ideas, especially the strategy of carefully creating rigorous definitions that capture intuitive notions. The hard work involved was carefully pinning down a subtle property, and one of the valuable skills that mathematicians learn is how to make vague ideas precise. Our investigation in this unit deepens the skill of abstraction that is one of the core strategies of mathematical inquiry. We began developing calculus by abstracting from a real life situation (well, almost): the physical scenario of Zeno shooting an arrow approaching a target. Here are the general inquiry skills we used while developing a precise notion of "approaching."

- **Selection of the Relevant Mathematical Objects** — The first step is deciding what mathematical objects we should study while pinning down the motivating concept. For us this process led at first to identifying sequences when studying convergence and then functions when studying continuity and differentiability. (pp. 91, 113)

- **Articulation** — The next step is to try to pin down our current understanding of the concept we seek to specify, in this case the

4.10. From Vague to Precise

idea of "approaching". This clarifying step includes addressing as many of the known ambiguities, unresolved conceptual conflicts, and subtleties as possible. Addressing these issues consists of producing a working definition. Sometimes we cannot articulate a full definition, so we instead produce necessary or sufficient properties along the way. (pp. 92–94)

[margin note: process of moving for the students not as a finished product handed down from a teacher]

- **Experimentation on Provisional Definitions** — After formulating a working definition, we move into a phase of testing. We look for examples that intuitively should have the property (in our case, the property of "approaching") and check whether they satisfy our working definition. We look for examples that intuitively should not have the property and check whether they do not meet our working definition. And we explain any deviations between the working definition and our examples explicitly. We also try to produce new examples with fewer special properties, working towards a collection of examples that can be considered general. This process makes us aware of new issues with our working definition and possibly new objects or concepts that need attention (like the tail of a sequence). (pp. 94–96)

- **Iteration** — And finally, we cycle back from testing our working definition to building a new definition that addresses visible issues. Sometimes we can only address some of the issues we are aware of, and sometimes we resolve an issue at the cost of creating a new one. We repeat this iterative process until we cannot identify any problematic issues with our working definition and we cannot produce any new examples that elicit fresh issues. Then the working definition becomes the definition. (pp. 96–113)

The exploration of convergence and limit was, of course, merely the tip of the iceberg. After pinning down the definitions of convergence and limit we could then explore the worlds of the derivative, the integral, and the many applications that they allowed us to understand. This experience allowed us to take a vague idea and turn it into a whole world of differentiated concepts. The march of idea after idea took us to continuity, to the derivative, and then to the integral.

In each case, part of our potent strategy of developing ideas and exploring them involved repeatedly clarifying our thinking until hazy intuition was transformed into formal definitions and clear thought.

5
Conclusion

5.1 Distilling Ideas

This textbook series includes several examples of mathematical investigation that illustrate a potent process of distilling ideas and then exploring them. They each present an example of mathematical inquiry and the rich consequences of its application.

- The investigation of graph theory begins with the famous Königsberg Bridge Problem. One of the first and foremost strategies of mathematical creation is the process of abstraction. From that famous problem we abstract the essential ingredients. And we see a whole collection of other questions and challenges whose fundamental issues are the same, namely, objects and connections. We are able to capture those essential elements in the definition of a graph. From there, we explore the definition by creating more examples and by identifying properties that any graph must have. We then distill ideas that allow us to pose our motivating problems in the abstract setting of graph theory. We solve our original problems and then undertake the process of extending our ideas further. As with each of our topics, the further exploration and application of graph theory can literally last a lifetime and can be of enormous importance.

- Our investigation of group theory arises from seeing commonalities among several familiar ideas such as the arithmetic operations and moving a block. By highlighting the essential similarities among those familiar activities, we produce the definition of a group. Once we have the definition of a group, we explore the definition by creating examples and seeing how we could generate more examples from old examples. We then explore the consequences of the definition of a group to see what properties must be true for all groups and what special features could distinguish some groups from others. One of the basic mathematical ideas is to describe functions that relate instances of structures that we have created. This impulse leads us to describe group homomorphisms and, in capturing sameness, group isomorphisms. The next step is to move toward classifying groups in some reasonable way.

- Calculus presents us with the challenge of taking an intuitive idea, namely, a moving object "approaching" its destination, and turning that intuitive idea into a mathematically rigorous definition. Historically, that challenge took mathematicians nearly 200 years. The method we demonstrate in the chapter entails a process of making many failed attempts and incrementally using those mistakes as an aid toward eventually creating an effective definition. We find features, such as the tail of a sequence, that are necessary concepts to employ before a good definition of convergence and limit can be constructed. We then recognize various different types of sequences and functions as we explore variations of the ideas we had pinned down.

- The exploration of number theory involves looking at the familiar integers and seeing relationships among them such as divisibility and identifying elemental building blocks such as primes. Looking at a cyclic version of the integers extends the concept of integers to create modular arithmetic. One of the guiding questions in exploring modular arithmetic is to seek similarities and differences between how the standard integers behave compared to how integers behave in modular arithmetic. Looking at examples and looking for patterns allows us to find unexpected relationships that

5.1. Distilling Ideas

have interesting consequences, some practical such as cryptography. Number theory shows us that the deliberate investigation of what might appear simple and easy can lead to whole worlds of previously unseen beauty and structure.

Many of the themes of learning from examples and isolating essential ingredients appear in all units in this series. In each case, we make progress toward developing mathematics by isolating ideas, distilling definitions, and then exploring consequences of those essential ideas and definitions.

The strategies of developing ideas that we see illustrated in these units serve us well as mathematicians. We look deeply at simple and familiar objects with the goals of identifying essential features and commonalities. Definitions arise as a result of this potent process of distilling ideas, finding consequences and connections, and continually refining vagueness into ever increasing precision. As you go on to learn more mathematics, you will see these themes and these strategies of investigation arising time after time. Soon these strategies will become a habitual part of how you think. As you continue your study of mathematics, these strategies of thinking will become a deeper and more meaningful part of who you are.

During your study with this textbook series, you have probably experienced several shifts in how you operate as a mathematician. You have gained mathematical maturity by becoming much more proficient at proving things on your own. You now know the power of returning to the definition rather than being vague. You have developed an attitude of specificity. You know how to decompose a statement, that is, you know how to identify the hypothesis and conclusion, how to frame negations and contradictions, how to deal with quantifiers. You have naturally produced proofs of several types including direct proofs, proofs by contradiction, inductive proofs, and others, and you recognize all of these as merely being examples of clear logical reasoning rather than some magical formulations. Concepts that at first seemed impossible to grasp turned into ideas that are second nature to you. You have learned how to explore and create mathematics.

By adding these skills to your arsenal of ways to investigate the world, you can apply those strategies in any arena. You have faced a collection of difficult and abstract ideas and found your own strengths to allow you to surmount difficulties and clarify vagueness. These examples in mathematics actually illustrate strategies of thinking that are applicable and have huge impact outside of mathematics as well. We hope that through this exploration, you have empowered yourself to grapple more effectively with all the unknowns and conundrums of life. In short, you have become more *mathematikos*, "inclined to learn."

Annotated Index

Mathematics relies on precise definitions, and much of the work of creating mathematics is putting an idea into a precise form. As a result, you should work with the definitions embedded in each inquiry unit. So, rather than containing the precise definitions, this Glossary gives reminders of the terms and links to their precise definitions in the units. The Glossary also discusses those terms that are used but not defined within the units; these terms are starred*.

abelian (groups: 60)
: A group is abelian if its binary operation is commutative.

action (groups: 82)
: A group action lets us think of a group as a set of permutations of some set.

adjacent (graphs: 13)
: Two vertices are adjacent if they are the endpoints of a common edge.

anti-derivative (calculus: 142)
: An anti-derivative of a function $f(x)$ is a function F such that $F'(x) = f(x)$.

automorphism (graphs: 33, groups: 81)
: An automorphism is an isomorphism of an object with itself, possibly in a non-trivial fashion.

bijective*, bijection* (graphs: 33, groups: 67,68,72)
: A function, $f : D \to C$, is bijective if (i) every element of the domain D is paired with a distinct element of the codomain C and (ii) every

153

element of the codomain C is paired with an element of the domain. In other words, a function is bijective if it is injective (property i) and surjective (property ii); less formally, a function is bijective (or a bijection) if every possible output is hit exactly once. A bijection is a correspondence between two sets that tells us how the elements are paired.

binary operation* (groups: 44)
A binary operation is a function with two inputs from a set, S, that gives an output in the set S. The definition of a function implies that the operation must be closed, meaning that the output is an element of S. The definition of a function also implies that the operation must be well-defined, meaning that for every pair of inputs, there is exactly one result; in other words, if the elements of S can be represented in multiple ways, then replacing an input in a binary operation with another representation does not affect the output.

bounded (calculus: 106)
A sequence or function is bounded if its values stay within a finite range. A sequence/function is bounded above if its values never get larger than a fixed value and is bounded below if its values never get smaller than a fixed value.

Cancellation Law (groups: 51)
In a group, $ax = ay$ if and only if $x = y$.

Cauchy sequence (calculus: 112)
A sequence is Cauchy if eventually every pair of terms in a tail are close.

Cauchy's Theorem (groups: 84)
If p is prime and p divides $|G|$, then G has an element of order p.

center (groups: 61)
The center of the group G, written $Z(G)$, is the set of elements that commute with all of the elements of G.

centralizer (groups: 85)
The centralizer of an element is the collection of elements that commute with it. This is a special case of a stabilizer.

circuit (graphs: 19)
A circuit is a walk that starts and ends at the same vertex and does not repeat edges.

Annotated Index 155

closed*
: see binary operation

coloring (graphs: 37)
: A coloring of a graph is an assignment of a color to every vertex such that adjacent vertices have different colors. An n-coloring is a coloring that uses at most n colors. A graph is n-colorable if it has an n-coloring.

commutative (groups: 48)
: A binary operation is commutative if the order of the inputs does not affect the result. The classic example is addition, $m + n = n + m$ for any values of m and n.

component (graphs: 29)
: A component is a whole connected piece of a graph.

congruent (groups: 53)
: Two integers, a and b, are congruent modulo n if n divides $b - a$, or equivalently if there is an integer k such that $b - a = nk$.

conjugacy class (groups: 85)
: The conjugacy class of an element is the set of elements that are conjugate to it. This is a special case of an orbit.

conjugation (groups: 74)
: Conjugation is a function from a group to itself. To conjugate h, by g, simply compute ghg^{-1}.

connected (graphs: 19)
: A pair of vertices is connected if there exists a walk between them. A graph is connected if all pairs of vertices in the graph are connected. Otherwise, the graph is disconnected.

continuous (calculus: 119)
: A function is continuous if points close together in the domain are sent close together in the codomain.

converge (calculus: 97)
: A sequence converges to ℓ if eventually all of the terms get within any specified distance from ℓ. A sequence that does not converge is said to diverge.

coset (groups: 62)
: Given a subgroup H and a group element g, the left coset gH is formed by starring every element of H on the left by g. Right cosets are built similarly with starring on the right.

cyclic (groups: 59)
: A group is cyclic if it can be generated by a single element.

decreasing (calculus: 106)
: see monotonic

definite integral (calculus: 136)
: The definite integral of $f(x)$ from a to b is the limit of approximations of the area between the graph of $f(x)$ and the x-axis. There are many other interpretations of this value.

degree (graphs: 13)
: The degree of a vertex v is the number of edges with v as an endpoint, where loops at v contribute twice.

derivative, differentiable (calculus: 128)
: A function is differentiable if its graph looks locally like a straight line. The derivative of a function $f(x)$ at x_0 is the slope of the line that best approximates the function near x_0. The derivative of a function is another function whose values give the slope of the graph of the original function at each point.

difference quotient (calculus: 127)
: The difference quotient is the slope of a secant line, meaning the change in the value of the function divided by the change in the inputs.

dihedral groups (groups: 55)
: The dihedral group, D_n, is the group of symmetries of the regular n-gon.

disconnected (graphs: 19)
: see connected

distance (calculus: 92)
: If two points x, y live in $\mathbb{R}$, the distance between them is just the absolute value of their difference, $\|y - x\|$.

diverge
: see converge

Annotated Index

dual (graphs: 36)
: Given a planar drawing of a graph, its dual is built by placing a vertex inside each face of G and adding edges between the vertices that come from faces that share a boundary edge.

edge (graphs: 10)
: see graph

endpoint (graphs: 13)
: The endpoints of the edge $\{v, w\}$ are the vertices v and w.

equivalence relation (graphs: 19, groups: 74)
: An equivalence relation captures the notion that a collection of objects can be partitioned into sets that are all "the same". For example, the set of ratios of integers with non-zero denominators. We say that $\frac{a}{b}$ is the same fraction as $\frac{c}{d}$ when $ad = bc$, written $\frac{a}{b} \sim \frac{c}{d}$. This allows us to group the ratios into sets that all represent the same number. There are three key components to an equivalence relation. (1) Reflexivity: For any ratio $\frac{a}{b}$, we can show that $\frac{a}{b} \sim \frac{a}{b}$. (2) Symmetry: If we know that $\frac{a}{b} \sim \frac{c}{d}$, then we can show that $\frac{c}{d} \sim \frac{a}{b}$. (3) Transitivity: If we know that $\frac{a}{b} \sim \frac{c}{d}$ and that $\frac{c}{d} \sim \frac{e}{f}$, then we can show that $\frac{a}{b} \sim \frac{e}{f}$. These properties are not automatic; they must be proved.

epimorphism (groups: 72)
: An epimorphism is a homomorphism that is surjective.

Euler Characteristic (graphs: 29)
: A planar drawing of a graph with $|V|$ vertices, $|E|$ edges, and $|F|$ faces has Euler Characteristic $|V| - |E| + |F|$.

Euler circuit (graphs: 21)
: An Euler circuit is a walk that contains all of the vertices and edges in the graph without repeating edges and returns to its starting vertex.

Euler Circuit Theorem (graphs: 21)
: A connected graph has an Euler circuit if and only if every vertex has even degree.

Euler path (graphs: 22)
: An Euler path is a walk that contains all of the vertices and edges in the graph without repeating edges.

face (graphs: 27)
: A planar drawing of a graph breaks the plane into several regions, which we call faces.

finite*, infinite* (graphs: 10, groups: 60, calculus: 91)
A set is finite if its elements can be paired with the elements of the set $\{1, 2, \ldots, n\}$ for some natural number n. Otherwise, the set is infinite.

First Isomorphism Theorem (groups: 79)
For a homomorphism, $\phi : G \to H$, the image group, $\text{Im}(\phi)$ is isomorphic to $G/Ker(\phi)$.

Four Color Theorem (graphs: 40)
Any planar graph without loops can be colored with four or fewer colors.

function*
Given two sets, D and C, a function is a relation that pairs every element of D with some element of C. We think of the elements of D as the inputs to the function and the elements of C as the potential outputs. The set D is called the domain of the function, and the set C is called the codomain of the function. If the function is called f, then we indicate the domain and codomain by writing them as follows: $f : D \to C$. Note that not every element of C needs to have an element of D that is paired with it; those elements in C that are part of pairs are called the range of f. In a function, the elements of D are each associated to one and only one element of C, a property called well-defined.

Fundamental Theorem of Calculus (calculus: 142)
The definite integral of a continuous function f on the interval $[a, b]$ is equal to $F(b) - F(a)$ for any function F that is an anti-derivative of f.

Fundamental Theorem of Finitely Generated Abelian Groups (groups: 80)
Finitely generated abelian groups are isomorphic to the product of cyclic groups.

generated, generator (groups: 58)
The subgroup generated by g, written $\langle g \rangle$ is the set of all group elements that can be formed by combining g and g^{-1} arbitrarily. Similarly, the subgroup generated by S is the set of all group elements that can be formed by combinations of the elements of S and their inverses. The element g and the elements of S are then called generators for those subgroups.

graph (graphs: 10)
A graph $G = (V, E)$ is a collection of vertices V and a collection of edges E, which are unordered pairs of the vertices.

Annotated Index

group (groups: 49)
: A group is a set along with an associative binary operation such that there is an identity element and every element has an inverse.

homomorphism (groups: 69)
: A homomorphism is a morphism of groups; in other words, it is a function that preserves the binary operations. Saying $\phi : G \to H$ is a homomorphism means that $\phi(g_1 *_G g_2) = \phi(g_1) *_H \phi(g_2)$.

identity element (groups: 49)
: An element e is an identity (in a group) if $e * g = g = g * e$ for all elements g.

image (groups: 70)
: Given a function, the image of a set (that is a subset of the domain of that function) is the collection of outputs associated with those inputs. The image of the entire domain is called the range.

inclusion (groups: 70)
: If $A \subset B$, then the function that maps every element a from A to itself in B is called the inclusion map.

increasing (calculus: 106)
: see monotonic

index (groups: 63)
: The index of a subgroup is the number of distinct cosets it has.

infimum (calculus: 107)
: The infimum of a set of real numbers is its greatest lower bound.

injective* (groups: 67,72)
: A function is injective if no two elements from its domain are sent to the same value in the codomain. Sometimes this property is called "one-to-one" elsewhere, but one-to-one can be confused with the property called bijective. Usually, to show that a function f is injective, it helps to assume that $f(a) = f(b)$ and try to prove that $a = b$.

integral
: see definite integral

Intermediate Value Theorem (calculus: 121, 144)
: If $f : [a,b] \to \mathbb{R}$ is continuous, then for any M with $f(a) \leq M \leq f(b)$, there exists a $c \in [a,b]$ such that $f(c) = M$.

interval (calculus: 100)

An open interval, written (a, b) is the set of points between a and b but not including a or b. Be careful, because the notation (a, b) can also represent an ordered pair. The closed interval $[a, b]$ includes the points between a and b including a and b. The half-open intervals $[a, b)$ and $(a, b]$ each contain the interior points and the endpoint with the square bracket.

inverse (groups: 49)

Given a group G with identity e, an inverse of g is an element h such that $g * h = e = h * g$. Then h is written as g^{-1}.

isomorphism (graphs: 33, groups: 72)

An isomorphism is a function that completely identifies two objects, meaning that it describes a relabeling of the elements (a bijection) that preserves structure. An isomorphism of graphs relabels the vertices and edges in a way that preserves endpoints. An isomorphism of groups is a homomorphism that is also a bijection of the group elements.

kernel (groups: 71)

The kernel of a homomorphism is the preimage of the identity element from the codomain group.

Königsberg Bridge Problem (graphs: 5)

Is it possible to traverse all of the bridges of Königsberg exactly once, returning to the starting land mass?

Kuratowski's Theorem (graphs: 31)

A graph is planar unless it contains a copy of K_5 or $K_{3,3}$.

Lagrange's Theorem (groups: 63)

In a finite group, the order of a subgroup divides the order of the group.

leaf (graphs: 23)

A leaf is a vertex of degree 1.

limit (calculus: 102,117)

see converge or continuous

loop (graphs: 12)

A loop is an edge from a vertex to itself.

maximal subtree (graphs: 24)

A subtree is maximal if adding any more edges from the graph would add circuits.

Annotated Index

Mean Value Theorem (calculus: 135, 144)
: For any differentiable function on a closed interval, there is a point in the interval at which the derivative of the function is equal to its average rate of change over the whole interval.

monomorphism (groups: 72)
: A monomorphism is a homomorphism that is injective.

monotonic (calculus: 106)
: A sequence is increasing if each term is greater than or equal to the previous term. A sequence is decreasing if every term is less than or equal to the previous term. A sequence is monotonic if it is either increasing or decreasing.

morphism (graphs: 33)
: In general, a morphism is a function between two similar objects that preserves structure. A morphism of graphs maps the vertices and edges of one graph to those of another in such a way that endpoints are preserved. A morphism of groups is usually called a homomorphism.

multiple edges (graphs: 11)
: A graph $G = (V, E)$ has multiple edges if E contains two copies of the same edge. A single loop is not a multiple edge, but two loops at the same vertex is a multiple edge.

normal (groups: 76)
: A subgroup is normal if the left coset by each element equals the right coset by that element.

orbit (groups: 83)
: The orbit of an element under an action is the collection of other elements to which it can be sent by the group.

planar, plane (graphs: 25, groups: 55)
: The plane is $\mathbb{R}^2$, the set of ordered pairs of real numbers. A graph is planar if it can be drawn in the plane without edge crossings.

Platonic solids (graphs: 33)
: A Platonic Solid is a polyhedron whose faces are congruent regular polygons and all of whose vertices have the same degree.

preimage (groups: 70)
: Given a function, the primage of a set (that is a subset of the codomain of the function) is the set of input values whose outputs land in the set.

162 **Annotated Index**

product (groups: 64)
: The (cartesian) product of two sets A and B, written $A \times B$, is the set of ordered pairs with the first element from A and the second from B. The direct product of two groups $G \times H$ is their cartesian product along with the natural binary operation, to make it a group.

projection (groups: 72)
: In a product (cartesian or direct), there are natural functions that take an ordered pair and forget one of the coordinates. These functions are called projections.

regular planar graph (graphs: 32)
: A graph is a regular planar graph if it has a symmetric planar drawing such that every vertex has degree at least 3 and every face has at least 3 sides.

sequence (calculus: 92)
: A sequence is an ordered list, usually with items indexed by $\mathbb{N}$.

stabilizer (groups: 83)
: The stabilizer of an element is the set of elements in the group (action) that leave it fixed in place.

subdivision (graphs: 30)
: A subdivision of a graph is formed by adding a finite number of vertices in the middle of existing edges.

subgraph (graphs: 20)
: A subgraph is a graph that is a subset of another graph.

subgroup (groups: 56)
: A subgroup is a group that is a subset of another group.

subsequence (calculus: 109)
: A subsequence is a sequence that is a subset of another sequence.

subtree (graphs: 24)
: A subtree is a tree that is a subgraph.

supremum (calculus: 107)
: The supremum of a set is its least upper bound.

surjective* (groups: 67,72)
: A function is surjective exactly when every element in the codomain is the output associated with at least one input from the codomain. This property is sometimes called "onto" elsewhere.

Annotated Index 163

Sylow's Theorem (groups: 85)
: If p is prime and p^i divides $|G|$, then G has a subgroup of order p^i.

symmetric group (groups: 67)
: The symmetric group on a set is the set of bijections from that set to itself, made into a group using function composition. If that set has n elements, the symmetric group is written S_n.

symmetric planar drawing (graphs: 32)
: A planar drawing of a graph is symmetric if every vertex has the same degree and every face has the same number of sides.

symmetry (groups: 55)
: A symmetry is a rigid motion of an object that returns it to its original position as a set, though the individual parts may be in different locations. Some reflections and rotations are examples of symmetries.

tail (calculus: 96)
: A tail of a sequence is a sublist of the sequence that starts at one term and contains all of the later terms.

total degree (graphs: 13)
: The total degree of a graph is the sum of the degrees of its vertices.

Triangle Inequality (calculus: 93)
: The distance from x to y is smaller than or equal to the sum of the distance from x to z and the distance from z to y.

traceable (graphs: 16)
: A graph is traceable if you can draw it without picking up your pencil or repeating any edges. A graph is traceable returning to the start if it is traceable such that the pencil starts and ends at the same location.

tree (graphs: 23)
: A tree is a connected graph with no circuits.

trivial, non-trivial (groups: 57)
: A subgroup H of G is trivial if $H = \{e\}$ or if $H = G$. Otherwise H is non-trivial.

vertex (graphs: 10)
: see graph

walk (graphs: 18)

A walk, $W : v_0, e_1, v_1, e_1, v_2, \ldots, v_n$, is a finite list of alternating vertices and edges such that each edge e_j has endpoints v_{j-1} and v_j, the vertices before and after it in the list.

well-defined*

see binary operation

List of Symbols

Sets — Let A and B be sets; let x and y be numbers.

$a \in A$	the **element** a is in A
$A \subset B$	A is a **subset** of B, meaning B contains A (and they might be equal)
$A \subsetneq B$	A is a subset of B but not equal to B
$A \cup B$	the **union** of A and B, built by combining the elements of A and B
$A \setminus B$	the set produced by removing any elements of B from A
$A \times B$	$\{(a,b) | a \in A, b \in B\}$, the **cartesian product** of A and B
$|A|$	the **cardinality** (or size) of A, the number of elements in A
$\max\{A\}$	the biggest element or **maximum** of A (if it exists)
$\min\{A\}$	the smallest element or **minimum** of A (if it exists)
$\mathbb{Z}$	the **Integers**, the set of positive and negative whole numbers and 0
$\mathbb{N}$	the **Natural Numbers**, the set of positive whole numbers
$\mathbb{Q}$	the **Rational Numbers**, the set of number that can be expressed as ratios of integers with non-zero denominator
$\mathbb{R}$	the **Real Numbers**, the set of decimal numbers
$\mathbb{R}^2$	$\mathbb{R} \times \mathbb{R} = \{(r_1, r_2) | r_1, r_2 \in \mathbb{R}\}$, the **plane**
(x, y)	$\{r | r \in \mathbb{R}, x < r < y\}$, the **open interval** from x to y; OR an **ordered pair**
$[x, y]$	$\{r | r \in \mathbb{R}, x \leq r \leq y\}$, the **closed interval** from x to y

Functions — Let $f : D \to C$ be a function with domain D and codomain C. Let g be a function.

$f(x) = y, x \mapsto y$	the element x is **mapped to** the element y
$g \circ f$	g **composed** with f, sometimes read as "g after f"
$\text{Im}_f(S), f(S)$	$= \{c \in C \mid c = f(s) \text{ for some } s \in S\}$, the **image** of the set S under f
$\text{Preim}_f(T), f^{-1}(T)$	$= \{d \in D \mid f(d) \in T\}$, the **preimage** of the set T under f

List of Symbols

Groups — Let $(G, *)$ be a group with binary operation $*$. Let $g \in G$; let H be a group, let $n \in \mathbb{N}$.

e, e_G	the **identity** (of G)
g^{-1}	the **inverse** of g
D_n	the **dihedral group**, the symmetries of the regular n-gon
C_n	the **cyclic group** of order n
$a \equiv b \mod n$	a is **congruent** to b **modulo** n; n divides $(b - a)$
$[a]_n$	the **congruence class** of a modulo n
$\mathbb{Z}_n$	the set of congruence classes modulo n
$\oplus_n$	addition modulo n
$\oplus$	modular addition
$n\mathbb{Z}$	the integers that are multiples of n
g^n	the element g starred with itself n times
$\langle S \rangle$	the subgroup generated by the set S
$Z(G)$	the **center** of the group G
gH, Hg	the left/right **coset** of a subgroup H by an element g
$[G : H]$	the **index** of a subgroup H in the group G
$G \times H$	the **direct product** of groups G and H
$(1\ 3\ 2)$	the one-line or cycle notation for a **permutation**
S_n	the **symmetric group** on n objects
$Sym(X)$	the **symmetric group** on the set X
$Ker(\phi)$	the **kernel** of the function ϕ
$G \cong H$	the group G is **isomorphic** to the group H
ϕ_g	**conjugation** by an element g
$K \triangleleft G$	K is a **normal** subgroup of G
G/H	the **quotient group** of G by a normal subgroup H, read "G mod H"
$[G, G]$	$\langle \{ghg^{-1}h^{-1} | g, h \in G\} \rangle$, the **commutator** subgroup of G
$SL_2(\mathbb{R})$	2-by-2 matrices with real entries and determinant 1
$Aut(G)$	the set of **automorphisms** of a group G
$Stab(x)$	the **stabilizer** of the element x
$Orb(x)$	the **orbit** of the element x

Graphs — Let G be a graph with vertices V and edges E.

$e = \{v_1, v_2\}$	an **edge** e with **endpoints** v_1 and v_2
$\deg(v)$	the **degree** of a vertex v
$\deg_G(v)$	the **degree** of a vertex v in a graph G
$W : v_0 \ldots v_n$	a **walk** from vertex v_0 to vertex v_n
K_n	the **complete** graph on n vertices
$K_{m,n}$	the **complete bipartite** graph on $m+n$ vertices
H_S	the subgraph whose vertices are colored with colors in S

Calculus — Let $S = (a_1, a_2, a_3, \ldots)$ be a sequence, which can also be written as $(a_n | n \in \mathbb{N})$ or $(a_n)_{n \in \mathbb{N}}$. Let x and y be numbers and let A be a subset of $\mathbb{R}$. Let $f(x)$ be a differentiable function. Let g be a continuous function.

$\|x\|$	the **absolute value** of x
$\|y - x\|$	the **distance** from x to y
$M_\varepsilon, N_\varepsilon$	positive integers associated with a sequence and the number ε
$S \to \ell$	S **converges** to ℓ; ℓ is the **limit** of S
$\sup\{A\}$	the **least upper bound** or **supremum** of A (if it exists)
$\inf\{A\}$	the **greatest lower bound** or **infimum** of A (if it exists)
$\Delta(f, x_0)(h)$	$= \frac{f(x_0+h)-f(x_0)}{h}$, the **difference quotient**
$f'(x), \frac{d}{dx} f(x)$	the **derivative** of $f(x)$
$\int_a^b g(x)\,dx$	the **definite integral** of $g(x)$ from a to b
$\int g(x)\,dx$	the **anti-derivatives** of $g(x)$
$\sum_{k=1}^{n} a_k$	$= a_1 + a_2 + a_3 + \cdots + a_{n-1} + a_n$

About the Authors

Brian P. Katz is an Assistant Professor of Mathematics at Augustana College in Rock Island, Illinois. He received his BA from Williams College in 2003 with majors in mathematics, music, and chemistry and his PhD from The University of Texas at Austin in 2011, concentrating in algebraic geometry. While at UT Austin, Brian received the Frank Gerth III Graduate Excellence Award and Frank Gerth III Graduate Teaching Excellence Award from the Department of Mathematics. Brian is a Project NExT Fellow, supported by Harry Lucas, Jr and the Educational Advancement Foundation.

Brian is a member of the Inquiry-Based Learning community, which focuses on helping students to ask and explore mathematical questions for themselves. Brian has given many talks about mathematics, teaching, and technology at national conferences and workshops, including as a plenary speaker for the 2012 MAA IBL PREP workshop. In particular, Brian has written and spoken about ways to use student-written wikis to extend the power of IBL beyond the classroom. Brian has helped to organize the RL Moore Legacy Conference hosted by the Academy for Inquiry-Based Learning, and he serves as a mentor for new practitioners within that organization.

Brian is liberally educated and is passionate about engaging in the Liberal Arts. He sees mathematics at the core of this educational perspective: connecting to the deductive reasoning of philosophy, the structure of communication of linguistics, the abstract beauty of the arts, and all forms of critical writing and speaking. Outside of mathematics, Brian has offered an interdisciplinary first-year course called Mind and Meaning, and one of his mathematics courses has been designated as part of the core curriculum at Augustana College as a "Perspective on Human Values and Existence."

As an educator, Brian enjoys helping all students become clearer thinkers and communicators and empowering people to move from consuming to producing knowledge.

Michael Starbird is Professor of Mathematics and a University Distinguished Teaching Professor at The University of Texas at Austin. He received his B.A. degree from Pomona College and his Ph.D. in mathematics from the University of Wisconsin, Madison. He has been on the faculty of the Department of Mathematics of The University of Texas at Austin except for leaves including as a Visiting Member of the Institute for Advanced Study in Princeton, New Jersey, and as a member of the technical staff at the Jet Propulsion Laboratory in Pasadena, California. He served as Associate Dean in the College of Natural Sciences at UT from 1989 to 1997.

Starbird is a member of the Academy of Distinguished Teachers at The University of Texas of Austin and is an inaugural member of The University of Texas System Academy of Distinguished Teachers. He has won many teaching awards, including the 2007 Mathematical Association of America Deborah and Franklin Tepper Haimo National Award for Distinguished College or University Teaching of Mathematics; a Minnie Stevens Piper Professorship, which is awarded each year to 10 professors from any subject at any college or university in the state of Texas; the inaugural award of the Dad's Association Centennial Teaching Fellowship; the Excellence Award from the Eyes of Texas, twice; the President's Associates Teaching Excellence Award; the Jean Holloway Award for Teaching Excellence, which is the oldest teaching award at UT and is presented to one professor each year; the Chad Oliver Plan II Teaching Award, which is student-selected and awarded each year to one professor in the Plan II liberal arts honors program; and the Friar Society Centennial Teaching Fellowship, which is awarded to one professor at UT annually. He is an inaugural year Fellow of the AMS. In 1989, Professor Starbird was the UT Recreational Sports Super Racquets Champion.

Starbird's mathematical research is in the field of topology. He has served as a member-at-large of the Council of the American Mathematical Society and on the national education committees of both the American Mathematical Society and the Mathematical Association of America. He currently serves on the MAA's CUPM Committee and on its Steering Committee for the next CUPM Curriculum Guide. He directs UT's Inquiry Based Learning Project. He has given more than 200 invited lectures at colleges and universities throughout the country and more than 20 minicourses and workshops to mathematics teachers.

About the Authors

Starbird strives to present higher-level mathematics authentically to students and the general public and to teach thinking strategies that go beyond mathematics as well. With those goals in mind, he wrote, with co-author Edward B. Burger, *The Heart of Mathematics: An invitation to effective thinking* (now in its 4th edition), which won a 2001 Robert W. Hamilton Book Award. Burger and Starbird have also written a book that brings intriguing mathematical ideas to the public, entitled *Coincidences, Chaos, and All That Math Jazz: Making Light of Weighty Ideas*, published by W.W. Norton, 2005, and translated into eight foreign languages. In 2012, Burger and Starbird published *The 5 Elements of Effective Thinking*, which describes practical strategies for creating innovation and insight.

Starbird has produced five courses for The Teaching Company in their Great Courses Series: *Change and Motion: Calculus Made Clear* (1st edition, 2001, and 2nd edition, 2007); *Meaning from Data: Statistics Made Clear*, 2005; *What are the Chances? Probability Made Clear*, 2007; *Mathematics from the Visual World*, 2009; and, with collaborator Edward Burger, *The Joy of Thinking: The Beauty and Power of Classical Mathematical Ideas*, 2003. These courses bring an authentic understanding of significant ideas in mathematics to many people who are not necessarily mathematically oriented. Starbird loves to see real people find the intrigue and fascination that mathematical thinking can bring.

David Marshall, Edward Odell, and Michael Starbird wrote *Number Theory Through Inquiry*, which appeared in 2007 in the MAA's Textbook Series. *Number Theory Through Inquiry* and the new Katz-Starbird book *Distilling Ideas: An Introduction to Mathematical Thinking* are books in the newly created MAA Textbook Subseries called 'Mathematics Through Inquiry'. This subseries contains materials that foster an Inquiry-Based Learning strategy of instruction that encourages students to discover and develop mathematical ideas on their own.